D1579748

STATISTICS

for

SCIENCE PROJECTS

Sallie Powell

Tutor in Biology,
Park Lane College, Leeds

Hodder & Stoughton

A MEMBER OF THE HODDER HEADLINE GROUP

To Ken, for his encouragement.

British Library Cataloguing in Publication Data

Powell, Sallie
  Statistics for science projects
  1. Science – Statistical methods
  I. Title
  507.2

ISBN 0 340 664096

First published 1996

Impression number	10	9	8	7	6	5	4	3	2	1
Year	2000		1999		1998		1997		1996	

Typeset by Wearset, Boldon, Tyne and Wear.
Printed in Great Britain for Hodder & Stoughton Educational, a division of Hodder Headline Plc, 338 Euston Road, London NW1 3BH by Bath Press

Contents

00008114   509.2

Why use Statistics?

Why do you need to use statistics in your Science project? In your investigation it is only feasible to make a relatively small number of measurements. However carefully you make these, there is usually some experimental error. You can use statistical techniques to assess how certain you can be that these measurements are typical of a much larger sample. You can then go on to make predictions of trends or patterns in a large sample.

It's probably a good idea to begin by having a look at exactly what we mean by 'statistics'. We can then go on to discuss how you can use them in your projects.

What is statistics?

The science of statistics can be defined as 'the systematic collection and study of numerical facts'. 'Statistics' can also be used to mean the plural of 'statistic' – a single item of data.

When you have collected your data (or statistics), you can present them using graphs or charts to show any **trends** or **patterns**.

You will then need to assess the **reliability** of your data. However carefully you perform your experiment, the data you collect usually shows some variation. The more variation in your results, the less reliable they are. You will need to use statistical techniques to quantify this variation so that you can assess the reliability of your data. You can then carry out further statistical tests to determine how certain you can be of the validity of your conclusions.

We can show when you need to use statistical tests by looking at some experiments done by students.

Some examples of experiments needing statistical tests

Example **1A**	*Experiment on parachute descents*

METHOD

Yasmin compared the time taken for two parachutes of identical shape and materials to descend from the same height. One had a canopy of 0.25 m² and the other one had a canopy of 0.5 m². She found that the smaller one took 5.3 s to descend but the larger one took 8.9 s. She realised that this could be a chance difference; that is owing to uncontrolled variables such as wind or air temperature. She decided to try two statistical techniques to test the reliability of her results. She repeated the experiment five times and found the arithmetical mean for each parachute. That is the sum of the descent times for each parachute divided by the number of descents.

She found the sample range for each parachute; that is the difference between the highest and lowest reading.

She arranged her results in size classes of 0.5 s (see Table 1.1). She illustrated the results of her experiment in a histogram.

RESULTS

Table 1.1 Descent times of parachutes with different areas of canopy

Area of canopy/m^2	Descent time/s					Mean descent time/s	Sample range/s
	Reading						
	1	2	3	4	5		
0.25	5.3	5.7	5.1	5.2	5.4	5.3	0.6
0.50	8.9	8.7	8.5	9.4	9.0	8.9	0.9

(handwritten: 11 - 32%, 10.11%)

Tally of frequency of descent times

Size class of descent time/s	0.25 m^2 canopy	0.5 m^2 canopy
4.5–4.9		
5.0–5.4	IIII	
5.5–5.9	I	
6.0–6.4		
6.5–6.9		
7.0–7.4		
7.5–7.9		
8.0–8.4		
8.5–8.9		III
9.0–9.4		II
9.5–9.9		

CONCLUSIONS

Yasmin concluded that she was almost 100% confident that the slower descent of the larger parachute was owing to the larger area of the canopy and not due to chance. If she repeated her experiment one hundred times, she would expect the larger parachute to take longer to descend every time.

She based her conclusions on the following observations:
- a comparatively large difference between the two means.
- a small sample range for each parachute; i.e. little variation.
- no overlap of the descent times for each parachute.

Example 1B   *Experiment on effect of cooking on vitamin C*

METHOD

The results of an experiment done by Gavin were not so clear cut. He compared the destruction of vitamin C in cabbage in boiling water and by microwave cooking. He took five samples in each type of cooking.

RESULTS

He found that the percentage loss in boiling varied from 16.1% to 18.9% with a mean of 16.8%. In the microwave cooking the range was from 14.3% to 16.9% with a mean of 15.4%. There was an overlap between his results with a small difference in the means (see Fig. 1.1).

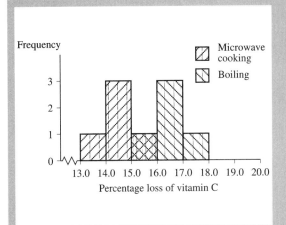

Figure 1.1 *Histogram of percentage loss of vitamin C in boiling and microwave cooking.*

If this small sample was truly representative, he could expect a frequency histogram of a larger sample of 100 to show a little overlap (see Fig. 1.2).

If his initial sample was not truly representative, a further 95 samples might show completely overlapping histograms. Each type of cooking might look like a single bell-shaped curve with all the readings clustered round a point (see Fig. 1.3). This would indicate that the difference between the means in his first sample could be due to chance – uncontrolled variables – rather than the method of cooking.

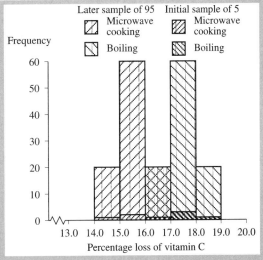

Figure 1.2 *Predicted histogram of percentage loss of vitamin C with a further 95 readings.*

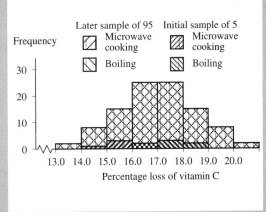

Figure 1.3 *Possible histogram of percentage loss of vitamin C with a further 95 readings.*

CONCLUSIONS

He decided to take a larger sample and carry out further statistical tests to estimate the amount of overlap he could expect in a large sample. He could then say how confident he could be that boiling destroyed more vitamin C than microwave cooking.

Example 1C — Experiment to compare insulating properties of granules and beads

METHOD

There was no overlap between the two sets of data in Suresh's experiment to compare the insulating properties of polystyrene granules and polystyrene beads. He took five **replicates** (repeated measurements) for each type of material.

He found that it took between 300 s and 308 s for water to cool from 80 °C to 70 °C with a mean of

306 s with granules, compared with 309 s to 313 s with a mean of 310 s with the beads.

This small difference between the means suggested that a larger sample might show an overlap. Suresh decided to carry out more replicates and use a statistical test to measure the confidence level of his conclusions.

Example 1D — Effect of caffeine on heart rate

METHOD

Caroline was investigating the effect of caffeine on heart rate by giving one group of ten students a cup of caffeinated coffee and another group of ten a cup of decaffeinated coffee. She took their heart rate every ten minutes.

RESULTS

After 30 minutes she found the heart rates of the caffeinated coffee group had a sample range showing an increase from 4.5% to 19.7%. The mean increase was 12.2%. The decaffeinated group showed a change ranging from a decrease of 2.4% to an increase of 2.1%. The mean change was an increase of 1.3%.

CONCLUSIONS

Although there was a considerable difference between the means of the two groups and no overlap of the data, the large sample range

suggested that a larger sample might show an overlap. Caroline could not be 100% confident that the caffeine was responsible for the difference. She took a larger sample and carried out statistical tests to measure the variation in her results and the confidence level of her conclusions.

The previous experiments have been looking for a *difference* between two sets of data. Let's have a look at a different type of experiment; one which is looking for a *correlation* between two sets of data.

Correlation means that as one variable changes, the other variable changes, e.g. as the temperature of the water increases, the heart rate of *Daphnia* increases. This does not necessarily mean that the rise in temperature has caused the heart rate to increase. It might be some other factor such as a decrease in the amount of oxygen dissolved in the water.

Example 1E — Experiment on chlorophyll content of leaves and iron

METHOD

Deema designed an experiment to see if there was an association between two sets of data. She wanted to know if there was any correlation between the amount of chlorophyll in tomato leaves and the concentration of iron salts in the culture solution they were growing in. She prepared a series of culture solutions with the concentration of iron salts varying from zero to

1 g dm^{-3}. She grew the tomato seedlings in the culture solutions for six weeks. She then extracted the chlorophyll from the leaves and measured the concentration by its absorbance in a colorimeter. She then looked for an association between the two sets of data by plotting a **scattergram** of her data (see Fig. 1.4).

RESULTS

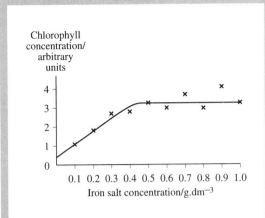

Figure 1.4 *Scattergram to show correlation between chlorophyll concentration of tomato leaves and iron content of the culture solution.*

CONCLUSIONS

Deema thought there was some **positive correlation** between the lower concentrations of the iron salts and chlorophyll content as a line drawn through the points inclines upwards. That is, the chlorophyll concentration increases as the concentration of iron salts increases. As all the points do not lie on this line, she could not be 100% confident that there was a correlation. She decided to carry out a statistical test to measure the confidence level for these low concentrations. At the higher concentrations of iron salts, the concentration of chlorophyll seemed to be more or less constant as the line is horizontal.

Causes of variation

In all these experiments, you can see that you would not expect to get exactly the same results every time you repeated the experiment. Let's have a look at some suggestions as to the causes of this variation.

- Variation in the environment during the experiment. There might be slight changes in the temperature of the laboratory which would affect the cooling experiment.
- Limitations of the apparatus. Laboratory thermometers might only be accurate to within 0.5 °C and stopwatches to within 0.5 s. The accuracy of the burette in the titration of vitamin C experiment might be limited to 0.1 cm³.
- Human error. The colour change in the end point of a titration might not be clear to some students. The time between the heat treatment and the titration might vary according to the skill of the experimenter. The early readings of the parachute descent might be less accurate as the student is less practised in the technique.
- Genetic variation in living organisms. In the chlorophyll experiment, the tomato plants might vary genetically in their ability to produce chlorophyll. In the caffeine experiment some people might be genetically more susceptible to caffeine than others.
- Age of the organisms. Older people might be less affected by caffeine.
- Previous environment of the organisms. Some people might be constant coffee drinkers and be more tolerant of its effects than others.

What variables would you control in the experiment to compare the effect of decaffeinated and caffeinated coffee on the students? The age, genetic variation and previous experience of the subjects in this experiment are difficult to control. Can you suggest any other variables which would be difficult to control? How would these variables affect the results?

Designing the experiment

When you design your experiment you must try to control these variables as far as possible to give reliable results. To allow for any uncontrolled variables, you will probably decide to repeat the experiment several times. These repeated measurements are called **replicates**. The set of replicates is called the **sample**. The larger the size of the sample, the more reliable the results of the experiment. The sample size will depend on:

- the nature of the experiment. A preliminary experiment will show how much variation you can expect in your results. This will give you some idea of the sample size you will need
- the time available
- the resources available
- the statistical techniques to be used.

This last point emphasises that the statistical tests must be decided upon during the planning of the experiment – not tagged on at the end.

Summary

You need to use statistics to predict trends and patterns in a large population from a small sample. You need to measure the variation in your data. The greater the variation, the less reliable your results. Statistical techniques are used:

- to show trends and patterns in your data
- to assess the reliability of your data
- to assess how certain you can be of the validity of your conclusions.

Statistical tests on your preliminary experiments such as finding the **arithmetical mean** and **sample range** will tell you if further statistical tests are necessary.

You will need to carry out further statistical tests if you are less than 100% certain of the validity of your conclusions. This would be the case in experiments looking for a difference if:

- the two sets of data overlapped
- in a small sample, there was a small difference between the means
- in a small sample, there was a large sample range.

In experiments looking for a correlation between two sets of data, statistical tests are needed to show the size of the correlation.

Sample is the set of measurements taken. The more uncontrolled variables in your experiment, the larger the size of the sample will need to be. Statistical tests must be decided on during the planning of the experiment.

chapter

2 Some Suggestions for Projects

The experiments discussed in Chapter 1 might have given you some ideas for investigations. There will be more experiments referred to in subsequent chapters. You might think of extending one of these or looking at a different angle on a theme. In the parachute experiment you could see if the shape or material of the canopy affects the descent time. If you enjoy cooking you might decide to compare the effect of a variety of cooking methods on the vitamin C content of the food. Here are some other suggestions for projects from different branches of science.

Projects in Biology

- Effect of nitrate, phosphate or potassium concentration on the growth of oat seedlings. Other factors affecting growth of seedlings you could study are light intensity, wavelength of light, temperature, salinity or acid rain.
- Effect of pollutants on the growth of duckweed.
- Effect of removing the flower buds on the growth of dandelions.
- Effect of exercise on resting pulse rate and recovery time.
- Comparing bacteria concentration in milk treated in various ways.
- Comparing the bacteriocidal effect of tears with antiseptics.
- The effect of caffeine on the rate of urine production.
- Effect of pectinase on fruit juice production.
- Relationship between lung capacity and smoking.
- Comparison of the flora in mown and unmown grass.
- Differential selection of leaves by earthworms.
- Environmental factors affecting leaf decomposition.
- Relationship between rate of water flow and species diversity in a stream.
- Monohybrid and dihybrid crosses in *Drosophila*.

Projects in Chemistry

- Comparing the bleaching power of various domestic bleaches.
- Comparing the effectiveness of indigestion tablets.
- Comparing the caffeine content of coffee and tea.
- Investigating the flammability of household materials.
- Comparing the percentage of salt in ready-salted crisps.
- Analysis of commercially available iron diet supplements.
- Comparing the efficacy of biological and non-biological detergents.
- Comparing the vitamin C content of different fruit juices.
- Comparison of sugar content of soft drinks.
- Factors affecting fractional distillation apparatus.
- Investigating the chemistry of swimming pools.

Projects in Physics

- Effect of temperature on the bounce of squash balls.
- Finding the force needed to deform ping pong balls.
- Effectiveness of friction brakes in motor vehicles.
- Strength of human hair.
- Strength of a fishing line under various conditions.
- The effectiveness of insulating materials against sound.
- Investigating the acoustics of a room.
- Assessing the performance of different audio cassettes.
- Investigation of the frequency range of a carbon granule microphone.
- Comparing the viscosity of different oils.
- Effect of temperature on the surface tension of various liquids.
- Absorption of microwaves by various materials.
- Effect of concentration on the refractive index of sodium chloride solution.

Projects in Psychology

- Comparing the difficulty of remembering sequences of numbers, letters and words.
- Comparing aural and visual memories.
- Effect of distracting tasks on the ability to memorise sequences.
- Comparing numerical and verbal ability in males and females.
- Effect of practice on IQ scores.
- To see if there is a correlation between an individual's speed on a manual task and word fluency.
- Comparing students' ability to solve problems with and without background music.
- To see if there is a relationship between interpersonal attraction and similar attitudes.
- To see if the colour of food affects the perception of its flavour.
- Investigation of the effect of age on reaction time.
- The effect of drugs such as caffeine on reaction time.

Points to bear in mind when choosing your project

Regulations of your examination board

Check the regulations carefully to make sure that your project complies with them. Most boards require a quantitative investigation involving measuring rather than just a description. Some boards allow students to work together in collecting the data providing they interpret them and write their projects up independently.

Interdisciplinary projects

Although these topics have been divided into different branches of science for convenience, the barriers between the disciplines tend to break down when tackling practical projects. Studies on biological washing powders or vitamin C could be tackled by either biology or chemistry students. Drugs and their effect on behaviour could be approached from a psychological or chemical angle.

Resources available

Some of these projects need a large number of subjects available. The effect of caffeine on heart rate would be easier to do during term time when a class of students could be persuaded to participate.

You will need to check that the apparatus and chemicals for the analytical experiments are available. Some apparatus could be improvised such as the parachutes in comparing descent times.

Time available

Check that you will be able to finish the project in the time available. Carrying out experiments tends to take you longer than you think as you often have to do preliminary experiments to perfect your technique. It is a good idea to make yourself an action plan setting out exactly when you are going to carry out each stage in your project.

It is probably best to restrict the design of the experimental work to two sets of measurements; that is comparing *two* sites or treatments. This will avoid the problem of the complex mathematics needed in tests involving *more than two* sites or treatments.

Safety

Make sure that you are aware of any safety precautions you need to take. Check 'Haz-cards' or similar safety instructions for any chemicals you use. Follow the instructions for aseptic techniques in microbiological experiments. Take appropriate precautions in experiments involving moving objects or electricity.

Time of the year

Experiments on growth of plants are more successful in the spring or summer. Ecological projects such as comparing the flora of mown and unmown grass also need to be done in the spring or summer when the species are easier to identify.

Originality

Although it is perfectly acceptable to take a project from a list, it is a good idea to try to think of a slightly different angle to take. You might be able to think of a problem you could study scientifically connected with the home or work.

Testing the claims made by advertisers for their products is a rich source of material! Whatever you do, try to choose a topic that really interests you.

Summary

There are some points to bear in mind in choosing a project:
- the requirements of your examination board
- the resources available to you

- time available
- safety precautions
- permission from appropriate authorities, e.g. access to land in ecological experiments; parents in measuring children
- time of the year must be suitable for ecological investigations
- originality
- a personal interest.

3 Choosing Statistical Tests

The most obvious way you would think of in handling the variation in your data is to find an average. Let's have a look at the ways of doing this.

Averages

There are three main ways of calculating an average: the **arithmetic mean**, the **median** and the **mode**.

Arithmetic mean is probably the most familiar way of measuring an average value. To find the mean pulse rate of a sample, the sum of the individual pulse rates is divided by the number of measurements in the sample. This is shown by the formula

$$\bar{x} = \frac{(\Sigma x)}{n}$$

where the mean of a sample is denoted by $\bar{x}$ (x with a horizontal line on top, pronounced x bar).

The measurements of the pulse rate are x.

The sum of the numbers is denoted by the Greek capital letter sigma Σ.

The sample size is denoted by n. e.g. the mean of the pulse rates 66, 82, 94, 73, 70, 75 beats per minute

$$= \frac{66 + 82 + 94 + 73 + 70 + 75}{6}$$

$= 76.66$ beats per minute.

Median is the value above which half of the measurements in the sample lie and below which the other half lie. When the size of the sample is an even number, arrange the numbers in ascending order. Take the pair of measurements in the middle of the sample and the median lies half way between these numbers. In the example above, the middle pair of numbers is 75 and 82. The median will be 78.5.

Medians are often useful where measurements are clustered towards one end of the scale. The measurements are said to have a **skewed** distribution.

An example of a skewed distribution is shown below in the height of 21 beech trees in metres, correct to one decimal place, in rank order (increasing order of height):

7.1, 7.3, 7.3, 8.4, 8.6, 9.1, 9.3, 9.5, 9.9, 9.9, 10.6, 10.6, 11.5, 12.7, 12.9, 13.3, 15.0, 17.4, 18.2, 18.6, 19.4.

The data was divided into equal size classes (see Table 3.1).

The skewed distribution can be shown in the histogram (see Fig. 3.1).

The median is 10.6 m which is lower than the mean of 11.74 m.

Table 3.1 Size classes of beech trees

Size class/m	Tally	Frequency
7.0–7.9	|||	3
8.0–8.9	||	2
9.0–9.9	ⅢⅡ	5
10.0–10.9	||	2
11.0–11.9			1
12.0–12.9	||	2
13.0–13.9			1
14.0–14.9	0	0
15.0–15.9			1
16.0–16.9	0	0
17.0–17.9			1
18.0–18.9	||	2
19.0–19.9			1

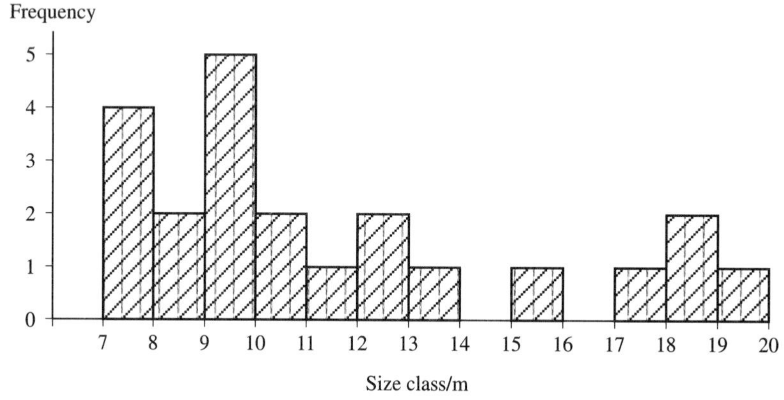

Figure 3.1 *Histogram of height of beech trees showing skewed distribution.*

The type of distribution is taken into account when deciding on a statistical test, as you will see later.

Mode is the measurement which occurs most frequently. The numbers of eggs per clutch in mallards were counted. The numbers were:

13, 9, 14, 11, 13, 10, 10, 12, 11, 8, 11, 11.

The mode is 11 eggs per clutch.

Sometimes there may be more than one modal value in a set of data. These are the pulse rates in beats per minute in a class of students:

68, 84, 76, 72, 60, 72, 82, 72, 78, 70, 69, 85, 78, 72, 81, 82, 78, 73, 74, 61, 78, 73, 75, 75.

The mode has two values: 72 and 78.

As well as finding an average value we need to know something about the other readings in our data. The simplest way of doing this is to find the sample range.

Sample range is the difference between the highest and lowest measurement in the sample. The sample range of the beech trees above is

$19.4 - 7.1 = 12.3$ m

The disadvantage of this statistic is that it depends on the highest and lowest measurements only and does not tell you anything about the clustering of the other measurements. This disadvantage is overcome with the second measure of scatter – sample standard deviation.

Sample standard deviation, s, is a way of measuring the spread of data around the sample mean. If your five measurements were 7, 8, 9, 10 and 11 m, if you calculated the average distance from the mean (9), the deviation would be:

$$\frac{-2 + -1 + 0 + 1 + 2}{5} = 0$$

To get rid of negative numbers, the deviations are squared and the average taken:

$$\frac{4 + 1 + 0 + 1 + 4}{5} = 2$$

In fact, in a small sample, it is usual to divide by one less than the sample size n; that is $(n - 1)$. In this example:

$$\frac{4 + 1 + 0 + 1 + 4}{4} = 2.5$$

This figure is called the **variance** s^2.

$$s^2 = \frac{\Sigma(x - \bar{x})^2}{(n - 1)}$$

The sample standard deviation (s) is the square root of the variance

$$s = \sqrt{\frac{\Sigma(x - \bar{x})^2}{(n - 1)}}$$

$$s = \sqrt{2.5}$$

$$s = 1.58$$

The **population standard deviation**, σ, is the standard deviation of the whole population; that is of all possible measurements or a very large sample. The formula for population standard deviation is:

$$\sigma = \sqrt{\frac{\Sigma(x - \bar{x})^2}{n}}$$

Sample and population
Sample is the limited set of measurements that are made, e.g. the heights of the beech trees that were measured.

Population is the larger number of measurements that it would be possible to take, e.g. the heights of all beech trees.

Example 3A | Calculating standard deviation

METHOD 1

We can work out the standard deviation in an experiment to compare the numerical ability of males and females. Each group was asked to count backwards in sevens from 300. The number of correct answers for each subject was recorded. The results for the males are recorded in Table 3.2.

Using the formula

$$s = \sqrt{\frac{\Sigma(x - \bar{x})^2}{n - 1}}$$

Table 3.2 Calculating standard deviation – method 1

Subject n	Score x	Score − mean $x - \bar{x}$	(Score − mean)^2 $(x - \bar{x})^2$
1	39	23.7	561.69
2	7	−8.3	68.89
3	12	−3.3	10.89
4	28	12.7	161.29
5	13	−2.3	5.29
6	23	7.7	59.29
7	18	2.7	7.29
8	12	−3.3	10.89
9	10	−5.3	28.09
10	13	−2.3	5.29
11	32	16.7	278.89
12	27	11.7	136.89
13	13	−2.3	5.29
14	8	−7.3	53.29
15	3	−12.3	151.29
16	19	3.7	13.69
17	15	−0.3	0.09
18	8	−7.3	53.29
19	12	−3.3	10.89
20	14	−1.3	1.69
21	14	−1.3	1.69
22	10	−5.3	28.09
23	7	−8.3	68.89
24	7	−8.3	68.89
25	14	−1.3	1.69
26	18	2.7	7.29
27	17	1.7	2.89
	$\Sigma x = 413$		$\Sigma(x - \bar{x})^2 = 1803.63$

1. Find mean value $\bar{x} = \dfrac{\sum x}{n}$ where $n = 27$.

$$\bar{x} = \frac{413}{27}$$

$$= 15.3$$

2. Find the difference between each score and the mean $(x - \bar{x})$.
3. Square the difference between each score and the mean $(x - \bar{x})^2$.
4. Find the sum of the squared differences between each score and the mean.

$$\sum(x - \bar{x})^2 = 1803.63$$

5. Divide by the size of the samples minus one.

$$\frac{\sum(x - \bar{x})^2}{n - 1} = \frac{1803.63}{26} = 69.37$$

6. Find the square root.

$$s = \sqrt{\frac{\sum(x - \bar{x})^2}{n - 1}}$$

$$= \sqrt{69.37}$$

$$= 8.329$$

METHOD 2

An alternative formula can be used to give a quicker method for calculating standard deviation.

$$s = \sqrt{\frac{\sum x^2 - ((\sum x)^2/n)}{n - 1}}$$

1. Complete the table of data by squaring the score of each subject.

Table 3.3 Calculating standard deviation – method 2

Subject n	Score x	Score squared x^2
1	39	1521
2	7	49
3	12	144
4	28	784
5	13	169
6	23	529
7	18	324
8	12	144
9	10	100
10	13	169
11	32	1024
12	27	729
13	13	169
14	8	64
15	3	9
16	19	361
17	15	225
18	8	64
19	12	144
20	14	196
21	14	196
22	10	100
23	7	49
24	7	49
25	14	196
26	18	324
27	17	289
	$\sum x = 413$	$\sum x^2 = 8121$

2. Find the sum of x and x^2.

$\sum x = 413 \qquad \sum x^2 = 8121$

3. Substitute the values of $\sum x$ and $\sum x^2$ into the equation for standard deviation.

$$s = \sqrt{\frac{\sum x^2 - ((\sum x)^2/n)}{n - 1}}$$

$$s = \sqrt{\frac{8121 - (413^2/27)}{26}}$$

$$s = \sqrt{\frac{8121 - (170569/27)}{26}}$$

$$s = \sqrt{\frac{8121 - 6317.37}{26}}$$

$$s = \sqrt{\frac{1803.63}{26}}$$

$$s = 8.329$$

Using your calculator to find standard deviation

You can find standard deviation on a scientific calculator. The procedure varies with different makes of calculators. We will take the Casio fx-570s as a typical example. You will need to consult the instructions booklet if you have a different make.

1. Switch the power on.
2. Find the statistics mode by pressing 'shift' and 'Scl'.
3. Set the function mode to 'SD' by pressing 'mode' and '2'.
4. Enter your data, e.g. 55, 54, 51, 55, 53, 53, 54, 52, by pressing the number

followed by 'M+':

55 M+; 54 M+; 51 M+; 55 M+; 53 M+; M+; 54 M+; 52 M+

5. Find sample standard deviation by pressing 'shift' and '$x\sigma n - 1$'. Sample standard deviation is 1.407885953.
6. Find population standard deviation by pressing 'shift' and '$x\sigma n$'. Population standard deviation is 1.316956719.

Standard deviation and the normal distribution

Some students studied the variation in the size of eggs produced on a poultry farm in a week. A total of 1056 eggs were laid. They weighed each egg, divided the masses into equal classes and constructed a histogram. It showed a *normal distribution*. Normal distribution is the distribution of the measurements of the population. This is shown on a histogram as being symmetrically distributed around a cluster in the centre, forming a bell-shaped curve (see Fig. 3.2).

In the case of a normal distribution, about 68% of all the measurements lie within one standard deviation of the mean (i.e. $\bar{x} \pm 1\,s$). That is if the mean is 76 g and the standard deviation is 4.5, 68% of the eggs fall in the range 71.5 to 80.5 g. About 95% of all values lie within two standard deviations of the mean (i.e. $\bar{x} \pm 2\,s$). That is 95% of the eggs fall in the range 67.0 to 85.0 g.

Standard error of difference

This is a measure of the difference between sample means of two sets of measurements.

You can calculate it using the formula:

$$SE_D = \sqrt{\frac{s_1^2}{n_1} + \frac{s_2^2}{n_2}}$$

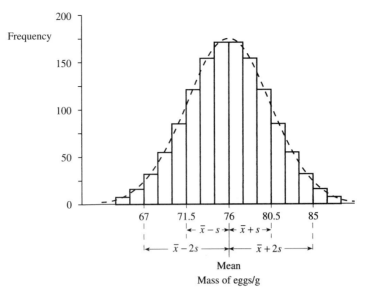

Figure 3.2 *Histogram of the mass of the eggs laid in a week on a poultry farm.*

SE_D is the standard error of difference.
s_1 is the standard deviation of the first sample.
s_2 is the standard deviation of the second sample.
n_1 is the size of the first sample.
n_2 is the size of the second sample.

Providing the size of each sample is greater than 30, by general agreement, if the difference between the arithmetic mean of the two sets of data is more than twice the standard error it is regarded as **significant**, i.e. due to factors other than chance. A difference of more than twice the standard error means roughly that there is a 5% probability that the difference is due to chance.

In Yasmin's parachute experiment, she and her friends timed 35 descents with a cotton parachute and 35 descents with an identical polythene parachute.

The mean for the cotton parachute was 5.4 s.

The mean for the polythene parachute was 5.6 s.

s_1 is 1.6. s_2 is 1.7.

$$SE_D = \sqrt{\frac{1.6^2}{35} + \frac{1.7^2}{35}}$$

$$= \sqrt{\frac{2.56}{35} + \frac{2.89}{35}}$$

$$= \sqrt{0.073 + 0.083}$$

$$= \sqrt{0.156}$$

$$= 0.395$$

Difference between the two means $= 5.6 - 5.4$ s
$\qquad\qquad\qquad\qquad\qquad\qquad = 0.2$ s.

This is less than twice SE_D, so we cannot be confident that the difference is significant.

Probability and confidence limits

In our experiments we aim to be able to find out how certain we can be that our results are due to the controlled variable and not due to chance.

Probability is the likelihood of an event occurring by chance. The probability of a simple event such as a tossed coin landing with 'heads' uppermost is $\frac{1}{2}$, alternatively expressed as 50% or 0.5.

$$\text{Probability} = \frac{\text{favourable outcome, i.e. heads uppermost}}{\text{total possible outcomes, i.e. heads or tails uppermost}}$$

If the event is 100% certain to occur, it has a value of 1. If it is impossible, it has a probability of 0. The probability, p, of an egg from the farm referred to in Fig. 3.2 weighing between 50 g and 90 g is 95% (i.e. $p = 0.95$). The probability of finding an egg outside this range is 0.05. (i.e. $p = 0.05$).

A confidence limit in data showing a normal distribution is defined as a given number of standard deviations either side of the mean. About 95% of the values will fall within a confidence limit of plus or minus two standard deviations of the mean.

Significance

As you can rarely be 100% sure that an event will occur, scientists have to decide what level they will accept. It is generally agreed that the 95% confidence level is acceptable. The probability of the event occurring by chance is 5% (i.e. $p = 0.5$). This is called the **significance level**.

Although 5% is the generally accepted significance level, you may use lower levels. There are various ways of expressing these probabilities of the results being due to chance:
- less than 1 in 20 is the 5% level or $p < 0.05$
- less than 1 in 100 is the 1% level or $p < 0.01$
- less than 1 in 1000 is the 0.1% or $p < 0.001$.

We will have a look at statistical tests you can do to measure the significance level of your results.

Finding significance levels

If an experiment involves collecting a number of measurements, you can use a statistical test to analyse the data and find the significance level of your conclusions. It is best to decide which test will be appropriate when designing the experiment to make sure that the right type and number of readings are taken.

If the results of an experiment are inconclusive, bigger samples might be needed. The same statistical tests can then be applied to the sample. These are some of the points to consider in choosing a statistical test.

The type of experiment
Are you investigating a *difference*, e.g. vitamin C breakdown with and without sodium hydrogen carbonate or an *association*, e.g. foot size in relation to height in humans? Or are you investigating something else?

The type of measurements
Are your measurements interval, ordinal or categorical (nominal)?

Interval data are measured in units where it is possible to say how much greater one measurement is than another, as in height.

Ordinal data are arranged in rank order, i.e. from smallest to largest. It is not possible to say exactly how much greater one measurement is than another, e.g. measuring sweetness as not sweet, slightly sweet, very sweet.

Categorical (nominal) data are data of different types but which cannot be arranged in order of size, e.g. brown mice and white mice in genetics.

Matched pairs or unmatched samples

Matched pairs are pairs of observations differing in one factor only, e.g. in the caffeine experiment the same person would drink caffeinated coffee and later drink decaffeinated coffee. Several subjects would repeat this.

Closely related subjects, such as identical twins, could be paired, one drinking caffeinated coffee and the other decaffeinated. One measurement from each of the samples can be paired with one, and only one, measurement from the other sample.

Using unmatched samples, one group of subjects would drink caffeinated coffee and another group would drink decaffeinated coffee. They might differ in other respects such as age, sex and the amount of caffeine they are used to. The two groups need not be the same size.

If possible, it is better to use matched pairs to give more reliable results.

Sample size

The sample size is the number of measurements taken.

Distribution of the measurements

You will need to do a preliminary experiment to give you some idea of the distribution of your data. One method of deciding on the distribution of the data is by comparing the mean, median and mode. If the data are normally distributed, then all three averages should be identical or very similar. Another method is to plot a histogram of the data to see if it shows a normal distribution or not.

When you have answered these questions, you will be able to use the flowchart shown overleaf to choose an appropriate test.

Carrying out the statistical test

Each statistical test has a formula which can be used to find the **test statistic**, e.g. t in the t-test; χ^2 in the chi-squared test. This test statistic enables you to decide if the difference or correlation you have found is great enough to be a genuine difference. That is, it is **statistically significant**. In the caffeine experiment you could conclude that the difference in heart rate was due to the caffeine. On the other hand, the difference might be so small that it is statistically insignificant and could well be due to chance.

Statistical tables are used to read off the probability of the difference as shown by the test statistic being due to chance. The critical level of probability is usually taken as 5% ($p = 0.05$). If the probability is below 5% (probability $p < 0.05$), then the difference is *not* due to chance but is great enough to be statistically significant.

If the probability is greater than 5%, the difference or correlation is so small that we cannot be confident that it is significant.

Choosing an appropriate statistical test

Is the experiment investigating
a **difference** or an **association**? *between each sites*
SA → light
intensity

Difference *between*
each site's SA.

Association *SA → light*

Are your measurements in
matched pairs with interval
data or unmatched samples?

Are your data in **categories,**
interval or **ordinal**?

categories

interval or
ordinal

Chi-squared test

Spearman rank
correlation
coefficient

7 to 30 samples;
scattered distribution

matched pairs
with interval data

unmatched samples

Ordinal or
interval data

Interval
data

Are your data
normally distributed?
Do you wish to test
for a difference
between **means**?
You can use this test
with a minimum of 6 to 10
measurements per
sample

t test for matched pairs

($t = z$ for larger numbers)

Do both sets of data
show **same-shaped**
distribution?
Do you wish to test for a
difference between **medians**?

One sample must be >1
and the other >4.

(use z if one sample >20)

Mann-Whitney *U* test

Ideally, each sample
>30? (If smaller,
t only approximate).

Are data **normally**
distributed?
Are the variances roughly
equal?
Do you wish to test
for a difference
between **means**?

t-test for
unmatched samples
($t = z$ for larger numbers)

Are the differences between
the matched measurements
from the two samples **symmetrically**
distributed?
Do you wish to test for a difference
between medians?
Do not use if the number of non-zero differences
is less than 5.

Wilcoxon matched-pairs test

We will have a look at an example of a *t*-test with unmatched samples in an experiment to compare the rate of decomposition of *Brassica* and *Bergenia* leaves in Chapter 6.

Summary

Averages

Arithmetic mean is the sum of the measurements divided by the number of measurements.

Median is the value above which half of the measurements in the sample lie and below which the other half lie.

Mode is the measurement which occurs most frequently.

Sample range is the difference between the highest and lowest measurements in the sample.

Standard deviation is a way of measuring the spread of data around the sample mean.

Standard error of difference is a measure of the difference between two sample means.

Probability is the likelihood of an event occurring by chance, denoted by *p*.

A **confidence limit** in data showing a normal distribution is defined as a given number of standard deviations either side of the mean.

Normal distribution is data symmetrically distributed round a cluster in the centre forming a bell-shaped curve.

Significance level is the probability of getting a result by chance.

Interval data are measured in units where it is possible to say how much greater one measurement is than another, e.g. mass.

Ordinal data is the arrangement of data in rank order but where it is not possible to say how much greater one measurement is than another, e.g. not sweet, slightly sweet, sweet or very sweet.

Categorical data are data of different types but which cannot be arranged in order of size, e.g. curly hair and straight hair.

Matched pairs is an experiment in which the same subject or a closely related subject is involved in each condition, e.g. the same *Daphnia* could have its heart rate measured at different temperatures. One measurement from one of the samples can be paired with one, and only one, measurement from another sample.

Unmatched samples is an experiment in which different subjects are involved in each condition, e.g. different *Daphnia* could be kept at different temperatures and their heart rate measured. There is no suggestion of pairing between the measurements of the two samples.

Statistical tests to find the significance level

Test	Type of experiment	Type of data	Matched or unmatched	Sample	Distribution
t-test for matched pairs	difference	interval	matched	6 to 10	normal
t-test for unmatched samples	difference	interval	unmatched	ideally >30	normal
Mann-Whitney U test	difference	ordinal or interval	unmatched	one sample >1; the other >4	same shaped
Wilcoxon matched-pairs test	difference	interval	matched	number of non-zero differences >5	symmetrical (not skewed)
χ^2	association	categorical	n/a	expected frequencies >5	n/a
Spearman rank correlation coefficient	association	interval or ordinal	n/a	at least 7 pairs	scattered distribution

4 Presentation of Data

When you have collected your data, you will need to present it as clearly as possible to show any trends or patterns. We will have a look at some of the ways you can do this.

Tables

Tables are used to show the data clearly in columns; each column has a heading with the units. We will have a look at a table of results from an experiment looking at the effect of temperature on the heart rate of the water flea – *Daphnia* (see Table 4.1).

Table 4.1 Heart rate of *Daphnia* at different temperatures

Temperature/ °C		2	10	20	30	40	50
Heart rate/ beats per minute	Reading 1	34	44	79	154	198	46
	Reading 2	30	49	88	160	210	43
	Reading 3	32	48	88	172	201	43
	Mean reading	32	47	85	162	203	44

Graphs

Graphs may be plotted to show any trends or relationships in your data clearly. It is important to select an appropriate type of graph to do this. This will depend on the type of data you collect; a set of data may be considered as being values of a **variable**.

A variable is a factor which changes (e.g. temperature, heart rate). An **independent variable** is chosen by the experimenter. In the *Daphnia* experiment, it is temperature. The **dependent variable** varies with the independent variable. In this case, it is the heart rate of the *Daphnia*.

The independent variable is always plotted along the horizontal axis (the *x*-axis). The dependent variable is plotted on the vertical axis (the *y*-axis).

Types of graphs

Line graphs are plotted where there is a gradual change in the variable. This type of variable is called a **continuous variable** as the measurement is not always a whole number. The heart rate of *Daphnia* might be 43.4 beats per minute at a particular temperature.

Line graphs are useful as intermediate readings can be found; heart rate of *Daphnia* at 15 °C can be estimated from the graph by **interpolation**. That is, raise a perpendicular from 15 °C on the *x*-axis to the graph line and then draw a perpendicular from the point of intersection to the *y*-axis (see Fig. 4.1a).

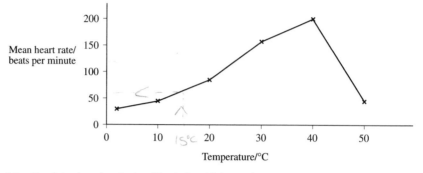

Figure 4.1a *Graph to show heart rate of* Daphnia *with increasing temperature.*

The graph line need not join the points: a **line of best fit** can be plotted to represent the results more realistically. Some of the points do not lie exactly on the line owing to experimental error. If a point is way out of line, make a note of this unexpected result (see Fig. 4.1b).

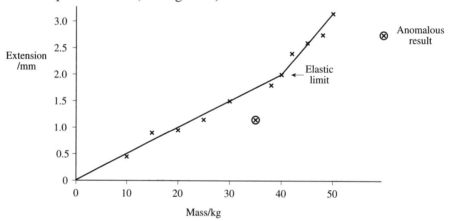

Figure 4.1b *Graph to show the extension of a loaded wire.*

Logarithmic graphs are useful in plotting growth rates which begin with small numbers increasing exponentially to very large numbers. The numbers of bacteria in raw milk kept at room temperature were estimated every hour for five hours (see Table 4.2).

Table 4.2 Population growth of bacteria in raw milk over 5 hours		
Time/hours	Numbers of bacteria/mm³	Logarithm of numbers of bacteria/mm³
0	230	2.362
1	480	2.681
2	1020	3.001
3	2200	3.342
4	4750	3.677
5	9500	3.978

A graph using arithmetic numbers shows the typical 'J-shaped' curve where there appears to be little change in numbers in the first hours. These data are not easy to handle as it is difficult to judge if the points lie on a curve (see Fig. 4.2).

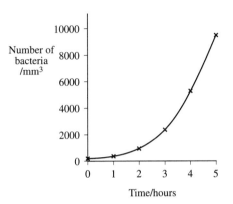

Figure 4.2 *Growth rate of bacteria in milk over 5 hours using arithmetic numbers.*

A logarithmic graph is easier to plot as the points lie on a straight line. You can look up the logarithm of each number of bacteria and plot the graph on ordinary graph paper (see Fig. 4.3).

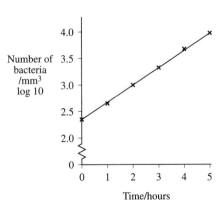

Figure 4.3 *Growth rate of bacteria in milk over 5 hours using the logarithm of the population numbers.*

It is more convenient to use semi-log graph paper to plot a logarithmic graph. This has intervals that appear equal but actually increase by a factor of ten on the *y*-axis. The *x*-axis has decimal ruling (see Fig. 4.4).

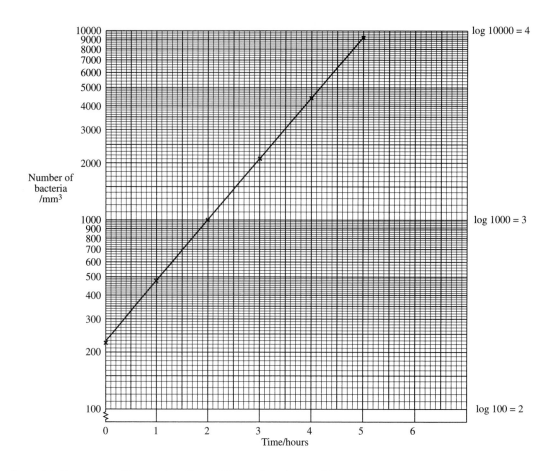

Figure 4.4 *Growth rate of bacteria in milk over 5 hours plotted on semi-log graph paper.*

Powers of ten

Powers of ten are used in scientific notation for very large or small numbers to avoid writing numerous zeros.

Number	Factor	Prefix	Sign
1 000	$\times 10^3$	kilo-	k
1 000 000	$\times 10^6$	mega-	M
1 000 000 000	$\times 10^9$	giga-	G
0.001	$\times 10^{-3}$	milli-	m
0.000001	$\times 10^{-6}$	micro-	µ

Points to note on plotting line graphs

- Title – e.g. heart rate of *Daphnia* against temperature.
- Correct axes – independent variable on *x*-axis, dependent on *y*-axis.
- Axes labelled with units e.g. heart rate/beats per minute. The sloping line (/) shows what is being measured to the left and units to the right.

- Use the largest possible scale. You can use a broken line to show a larger interval between zero and your first reading if necessary.
- Plot the points accurately with a '*x*' or '•' using a sharp pencil.
- Use a ruler to join points in a straight line or jagged line. Join points in a smooth curve if they are obviously in a curve. Some points might not lie exactly on the curve owing to experimental error. The line or curve of **best fit** is the line which represents the results most realistically allowing for experimental error (see Fig. 4.1b).
- If there is a point which is unexpected as it is way out of line with the others you might decide not to plot your line through it, but always include a note explaining this anomalous result (see Fig. 4.1b).
- Do not extrapolate the curve; that is, do not extend it beyond the readings you have taken – not above 50 °C or below 2 °C in the case of the *Daphnia* experiment.

Describing your graph

When you have plotted your graph, you will need to describe it – what is it showing? In the *Daphnia* experiment, the graph shows that the heart rate of *Daphnia* increases as the temperature increases up to a maximum at about 40 °C and then the heart rate decreases with a further increase in temperature.

In looking at the relationship between iron salt concentration and chlorophyll concentration of tomato leaves (see Fig. 1.4), you would note that the chlorophyll concentration of the tomato leaves increases as the iron salt concentration increases up to a maximum and then it begins to level off.

Bar graphs and histograms

These are used where dependent variables on the *y*-axis are whole numbers called **discrete variables**, as fractions are impossible. The number of students able to roll their tongue would be a discrete variable as you would not find half a student.

Bar graphs have separate bars on the graph to show that there are no intermediate conditions on the *x*-axis. Students in a class were asked to roll their tongues; most could do it and a few could not but no student could partly do it (see Fig. 4.5).

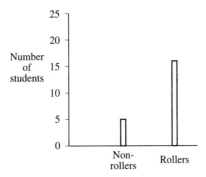

Figure 4.5 *Bar graph to show the number of students in a class able to roll or not roll their tongues.*

Histograms are used where there are intermediates on the *x*-axis. In measuring the height of students in a class, there were a range of intermediates between the two extremes.

Constructing a histogram

The following heights of students were recorded in metres:

1.66, 1.51, 1.64, 1.75, 1.73, 1.74, 1.60, 1.70, 1.66, 1.62, 1.58, 1.71, 1.75, 1.63, 1.68, 1.81, 1.54, 1.67, 1.77, 1.58, 1.69, 1.55, 1.72, 1.68, 1.61, 1.64, 1.57, 1.78, 1.78, 1.75

The heights are divided into eleven classes for convenience. The smallest height is subtracted from the largest one to find the range of heights. This is then divided by ten to find the class size.

Sample range $= 1.81 - 1.51$

$\qquad\qquad = 0.30$

Class size $\quad = \dfrac{0.30}{10}$

$\qquad\qquad = 0.03$ m

To avoid missing a height, cross out each height in turn and systematically enter the tally in the table (see Table 4.3). Add up the tally marks to find the frequency in each class. Construct a histogram of the height of the students. There are no spaces between the columns to show that the data are continuous (see Fig. 4.6).

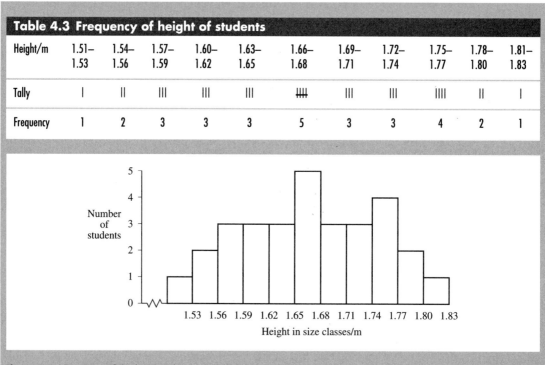

Table 4.3 Frequency of height of students

Height/m	1.51–1.53	1.54–1.56	1.57–1.59	1.60–1.62	1.63–1.65	1.66–1.68	1.69–1.71	1.72–1.74	1.75–1.77	1.78–1.80	1.81–1.83
Tally	I	II	III	III	III	IIII	III	III	IIII	II	I
Frequency	1	2	3	3	3	5	3	3	4	2	1

Figure 4.6 *Histogram of the height of a class of students.*

Kite diagrams are used to show the distribution of species in a habitat. A line called a transect was placed from a meadow across a marsh to a pond. A square 0.5×0.5 m called a quadrat was placed at 2 m intervals along the transect. The area covered by each species of *Ranunculus* was measured. The results are shown in a kite diagram (see Fig. 4.7).

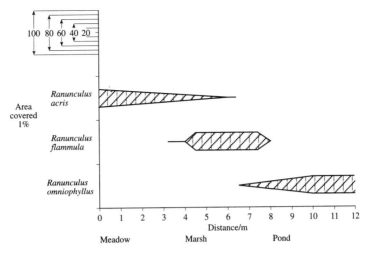

Figure 4.7 *Distribution of species of buttercup (*Ranunculus *sp.) from a meadow across a marsh to a pond.*

Scattergrams

Scattergrams are used to see if there is a correlation between two variables. The height and right foot length of each boy in a class of eight-year-olds were measured and recorded in a table (see Table 4.4).

Table 4.4 Table of height and right foot length of a class of eight-year-old boys

Boy	1	2	3	4	5	6	7	8	9	10	11	12
Height/cm	150	155	135	125	165	140	164	147	130	160	145	150
Foot length/cm	21.8	25.8	22.0	19.0	27.8	24.2	25.9	23.9	20.0	23.8	22.1	23.9

The scattergram (see Fig. 4.8) shows that there is a **positive correlation** between height and foot length as generally foot length increases as height increases.

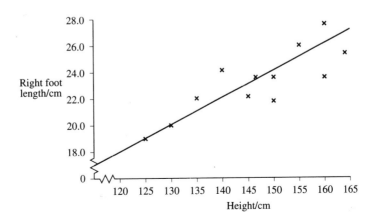

Figure 4.8 *Scattergram to show the relationship between the height of eight-year-old boys and their foot length.*

The mean mass of piglets in litters of different sizes was recorded (see Table 4.5).

Table 4.5 Table of litter size and mean size of piglets						
Litter number	1	2	3	4	5	6
Number of piglets in litter	3	7	10	8	6	9
Mean mass of piglet/g	1375	812	523	769	930	654

The scattergram of the average mass of piglets in the litter against litter size shows a negative correlation (see Fig. 4.9) as generally the mean mass of the piglets decreases as the litter size increases.

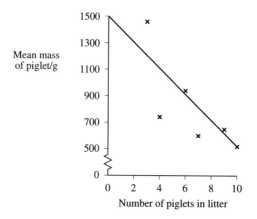

Figure 4.9 *Scattergram to show relationship between litter size and mean mass of piglets.*

A student plotted a scattergram showing the height of (human) fathers and their adult daughters. She found that the points were randomly scattered (see Fig. 4.10). This showed that there was no correlation between heights of fathers and their daughters.

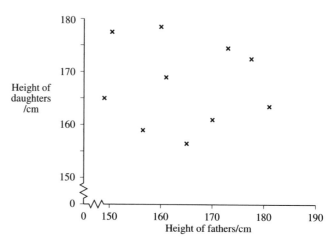

Figure 4.10 *Scattergram to show the relationship between the height of fathers with their daughters.*

Pie charts

These are useful in showing percentages or fractions of a whole. The percentage of the different ABO blood groups in the UK can be shown on a pie chart.

Summary

The **independent variable** is the variable chosen by the experimenter.

The **dependent variable** varies with the independent variable.

Tables are used to record raw data.

Line graphs are used when the dependent variable shows a continuous change with the independent variable. Fractions are possible in each variable.

Bar graphs are used when there are distinct categories in the independent variable with no intermediates (e.g. rollers or non-rollers of the tongue) and the dependent variable is measured in whole numbers (e.g. number of students).

Histograms are used when there are intermediates between the two extremes in the independent variable (e.g. all intermediates are possible between the two extremes in height of humans).

Kite diagrams are used to show a change in the distribution of species.

Scattergrams are used to show a correlation between two variables.

Pie charts show fractions of a whole.

Writing up your Project

You will need to write up your investigation in the style of a scientific paper. It's a good idea to have a look at an article in a scientific journal such as the *British Medical Journal* to use as a model. They are usually written up in the style 'a sample of thirty students was taken' but 'we took a sample of thirty students' is acceptable. You will see that the same sort of information is always presented with minor variations in the order and manner of presentation. We will have a look at a typical example. You can check with your examination board for their specific requirements.

Title

Put the title on the cover page with your name. Aim for something concise which will let your reader know at a glance the nub of your study. 'Sodium hydrogen carbonate and vitamin C' is incomplete. 'The effect of sodium hydrogen carbonate on vitamin C in cooking cabbage' is about right.

Contents

List the contents such as 'abstract' and 'introduction' with the page numbers.

Abstract

This is a summary in a single paragraph of about 200 words of the investigation. It should contain a sentence or two on the background and aim; an indication of the method; and should end with a concise statement of your results and conclusions.
- What were you investigating?
- What was your hypothesis?
- How did you investigate it?
- What statistical test did you use?
- What controls did you set up?
- How did you measure your results?
- What did you find?
- What conclusions did you draw? Was your hypothesis supported or rejected?

Results

Introduction

This puts the investigation into context. You will need to have read the relevant literature on your topic. It is sensible to begin by outlining your subject in general terms and then to go on to your specific investigation explaining why you have chosen to study it.

Conclude with a specific statement of the aims of your investigation.
- What is the chemical composition of vitamin C?
- What are its properties?
- How is it affected by heat?
- What else might affect it in cooking?
- What is sodium hydrogen carbonate? Why is it added to cabbage in cooking?
- What is the function of vitamin C in the body?
- What are the recommended daily amounts in the diet?
- What are the symptoms of a deficiency?
- What are the aims of your investigation?

Refer to the bibliography in your introduction to explain where you researched your information, e.g. recommended daily amount of vitamin C is 30 mg (4), where (4) is the reference number in the bibliography.

The bibliography should be at the end of your report listing reference number, author, title, journal or book, publisher, date and page number.

(4) Green R.P.O., Stout G.W. and Taylor, D.J., *Biological Science 1*, Cambridge University Press, 1984, p. 299.

Arrange your list of references in alphabetical order of the first author.

Hypothesis

State the prediction you are testing in a sentence as either the null hypothesis or the alternative hypothesis.

Null hypothesis

Sodium hydrogen carbonate has no significant effect on the breakdown of vitamin C in cooking cabbage.

Alternative hypothesis

Sodium hydrogen carbonate has a significant effect on the breakdown of vitamin C in cooking cabbage.

Method

Describe your method in sufficient detail for the reader to repeat the experiment in precisely the same way that you carried it out. Explain why you carried out each stage of the procedure. Emphasise safety precautions. Explain why you used a particular piece of apparatus or reagent.

In the vitamin C experiment:
- How did you design a 'matched pairs' experiment? Explain how you used the same cabbage for each treatment.
- How did you prepare your cabbage?
- How did you cook it? What apparatus did you use? What volume of water did you use? How long did you cook it for? What was the temperature during cooking?
- How much sodium hydrogen carbonate is added to the water? Why did you decide on this amount?

- How did you set up a control to your experiment? How did you make sure it was exactly like the experiment except for one factor – the presence of sodium hydrogen carbonate?
- How did you cool the cabbage after cooking? How long did this take?
- How did you extract the vitamin C?
- Why did you use orthophosphate? What volume and concentration did you use?
- Why did you titrate with DCPIP?
- What concentration of DCPIP did you use?
- What volume and concentration of ascorbic acid did you use in your standardisation?
- How did you do your standardisation? What was your end-point? How many times did you repeat the titration?
- How did you titrate your extracts of cabbage after cooking with and without sodium hydrogen carbonate? How many times did you repeat the titration?
- How did you control the effect of orthophosphate and sodium hydrogen carbonate on the DCPIP in the titration?
- How many samples of each treatment of cabbage did you take? Why did you choose this number? Refer to the statistical test you have chosen in selecting the sample size.

Results

Present your results clearly and concisely. Raw data and calculations in the statistical analysis should be in the appendix.

To make any trends or patterns clear, use appropriate tables, line graphs, bar charts, histograms or pie charts. These are discussed in Chapter 4.

Include headings for your tables and graphs. Tabulate the results of your statistical analysis.

Table 5.1 Effect of sodium hydrogen carbonate on vitamin C in cooking cabbage				
Mean concentration vitamin C in uncooked cabbage (mg/100 g)	Mean % loss in vitamin C after cooking with sodium hydrogen carbonate	Standard deviation	Mean % loss in vitamin C after cooking without sodium hydrogen carbonate	Standard deviation

Statistical analysis

$t = ?$ $p = ?$

If $p < 0.05$, the difference is statistically significant at the 5% significance level. State whether the difference is significant or not. A frequency histogram would be useful in illustrating the variation in your results.

Summarise the results in words.

Conclusions

Conclusions or inferences must be based only on the results obtained and clearly stated. State either that:

- the null hypothesis is supported or rejected or
- the alternative hypothesis is supported or rejected or
- rejected but accepted in a modified form.

The null hypothesis is that there is no significant difference between loss of vitamin C in cooking with or without sodium hydrogen carbonate at the 5% significance level.

The alternative hypothesis is that there is a significant difference in loss of vitamin C in cooking with sodium hydrogen carbonate at the 5% significance level. State if the loss is greater or less in the presence of sodium hydrogen carbonate.

Discussion

Discuss the **implications and relevance** of your conclusions. What is the effect of adding sodium hydrogen carbonate in cooking cabbage on the recommended daily amount of vitamin C in the diet?

Evaluate the limitations of the experiment – it was only tried on one cabbage.

Consider the reliability and sources of error:
- the cloudy extract tends to obscure the end point
- time between cooling the cabbage and titrating it varied
- shredding of cabbage was not uniform.

Suggest any modifications to the original design based on the evaluation of the results.

Appendix

Include all the raw data, calculations and statistical analysis.

Bibliography or footnotes

List your references in alphabetical order of author name, each reference in the form:

Reference number, author(s), title, journal or book, publisher, date, page.

See the Bibliography in this book as an example.

Acknowledgements

Indicating the source and extent of any help that has been received.

The last thing to do is to number the pages in your written account. You are usually asked to submit the investigation in a light paper folder for convenience, not a heavy ring binder.

Summary

Write up your project under the following headings:

Title – a single sentence explaining your project.

Introduction – background information, aims and hypothesis.

Method – an account of exactly how you carried out the experiment.

Results – tables, appropriate graphs or charts and a description of your results.

Conclusions – conclusions (there will probably just be one) based on the results. State the confidence level at which your hypothesis is accepted or rejected.

Discussion – an evaluation of your experiment discussing its limitations, sources of error, modifications and implications.

Appendix – raw data and calculations.

Bibliography – a list of references in alphabetical order.

Acknowlegements – any sources of help given.

The *t*-test for Unmatched Samples

- to test for differences in means of two sets of data.
- for two unmatched samples.
- for interval measurements.
- provided the data can be assumed to come from normally distributed populations with equal variances.
- ideally, your samples should have more than 30 measurements each. For smaller samples, the value of *t* might be only approximate.

Example 6A | *To compare the rate of decomposition of elephants' ears (Bergenia) and cabbage (Brassica) leaves*

METHOD

30 discs from each type of leaf were cut, each the size of a 2p piece. The area of the discs was measured. The discs were placed in rows alternately on the surface of a seed tray full of moist garden soil. 10 earthworms were added to the soil. The tray was placed in a black bin liner, loosely tied, and left on the laboratory bench. The discs were examined each week. After six weeks the area of each disc was measured using graph paper.

PRESENTATION OF DATA

Tabulate:
- the area of each disc after six weeks (x)
- the mean area of the discs of each species after six weeks ($\bar{x}$)
- the difference between the area and the mean area ($x - \bar{x}$)
- the difference between the area and the mean area squared ($x - \bar{x})^2$.

Table 6.1 Area of *Bergenia* leaves after decomposition

Disc number	Area after six weeks (mm²) x_1	Mean of area after six weeks (mm²) $\bar{x}_1$	Difference between area of disc and mean (mm²) $(x_1 - \bar{x}_1)$	Difference squared $(x_1 - \bar{x}_1)^2$
1	400	265	135	18225
2	250	265	−15	225
3	100	265	−165	27225
4	270	265	5	25
5	200	265	−65	4225
6	300	265	35	1225
7	300	265	35	1225
8	100	265	−165	27225
9	200	265	−65	4225
10	300	265	35	1225
11	200	265	−65	4225
12	240	265	−25	625

Table 6.1 continued

Disc number	Area after six weeks (mm²) x_1	Mean of area after six weeks (mm²) $\bar{x}_1$	Difference between area of disc and mean (mm²) $(x_1 - \bar{x}_1)$	Difference squared $(x_1 - \bar{x}_1)^2$
13	300	265	35	1225
14	250	265	−15	225
15	160	265	−105	11025
16	260	265	−5	25
17	300	265	35	1225
18	250	265	−15	225
19	260	265	−5	25
20	260	265	−5	25
21	280	265	15	225
22	240	265	−25	625
23	240	265	−25	625
24	150	265	−115	13225
25	260	265	−5	25
26	340	265	75	5625
27	350	265	85	7225
28	350	265	85	7225
29	420	265	155	24025
30	410	265	145	21025
	$\sum x_1 = 7940$			$\sum(x_1 - \bar{x}_1)^2 = 183\,750$

The initial area of each leaf disc was 500 mm².

Mean area of the discs $\bar{x}_1 = \dfrac{\sum x_1}{n_1}$

$= \dfrac{7940}{30}$

$= 265$ mm²

The initial area of each leaf disc was 500 mm².

Mean area of the discs $\bar{x}_2 = \dfrac{\sum x_2}{n_2}$

$= \dfrac{10210}{30}$

$= 340$ mm²

Table 6.2 Area of Brassica leaves after decomposition

Disc number	Area after six weeks (mm²) x_2	Mean of area after six weeks (mm²) $\bar{x}_2$	Difference between area of disc and mean (mm²) $(x_2 - \bar{x}_2)$	Difference squared $(x_2 - \bar{x}_2)^2$
1	400	340	60	3600
2	480	340	140	19600
3	475	340	135	18225
4	310	340	−30	900
5	330	340	−10	100
6	350	340	10	100
7	360	340	20	400
8	380	340	40	1600
9	360	340	20	400
10	350	340	10	100

Table 6.2 continued

Disc number	Area after six weeks (mm^2) x_2	Mean of area after six weeks (mm^2) $\bar{x}_2$	Difference between area of disc and mean (mm^2) $(x_2 - \bar{x}_2)$	Difference squared $(x_2 - \bar{x}_2)^2$
11	300	340	−40	1600
12	350	340	10	100
13	360	340	20	400
14	390	340	50	2500
15	210	340	−130	16900
16	160	340	−180	32400
17	210	340	−130	16900
18	320	340	−20	400
19	430	340	90	8100
20	360	340	20	400
21	440	340	100	10000
22	260	340	−80	6400
23	270	340	−70	4900
24	300	340	−40	1600
25	210	340	−130	16900
26	420	340	80	6400
27	440	340	100	10000
28	410	340	70	4900
29	415	340	75	5625
30	160	340	−180	32400

$$\sum x_2 = 10\ 210 \qquad \sum(x_2 - \bar{x}_2)^2 = 223\ 850$$

PRESENTATION OF DATA

The difference between the areas of the two sets of leaf discs can be shown by plotting the frequency of the final area of the discs on a histogram. To simplify the data, the measurements are divided into ten classes by dividing the difference between the largest and smallest final area by 10. This is the class size which is 50 mm².

The frequency of the areas in each class in each leaf can then be tabulated.

Table 6.3 To show the frequency of the leaf disc areas in classes after decomposition

Final area discs/mm²	0–49	50–99	100–149	150–199	200–249	250–299	300–349	350–399	400–449	450–499
Bergenia			||	||	₩₩ |	₩₩ ||||	₩₩ |	||	|||	
Frequency	0	0	2	2	6	9	6	2	3	0
Brassica				||	|||	||	₩₩	₩₩ ||||	₩₩ ||	||
Frequency	0	0	0	2	3	2	5	9	7	2

The frequency of the areas can be shown on a histogram using the same x-axis for both sets of data.

As the data for each set of leaf discs seem to be more or less symmetrically clustered round a central point, you can assume they are normally distributed. You can use a t-test to see if there is a significant difference between the rate of decomposition of the two species of leaves.

INTERPRETATION OF DATA

The results show that there is variation in the rate of decomposition of the discs of both species. The greater the variation, the less reliable the data. Standard deviation is a measure of this variation. It can be calculated as s_1 and s_2 from the formula below.

The standard deviation of the area of the *Bergenia* leaf discs after six weeks is 87.86 mm^2. The standard deviation of the area of the *Brassica* leaf discs after six weeks is 87.86 mm^2.

The histogram shows the overlap between the two sets of data. The greater the overlap, the less certain we can be that there is a significant difference between the two sets of data.

The *t*-test can be used to say just how certain we can be that there is a significant difference between the two sets of data.

FORMULA

$$t = \frac{|\bar{x}_1 - \bar{x}_2|}{\sqrt{\left(\dfrac{s_1^2}{n_1} + \dfrac{s_2^2}{n_2}\right)}}$$

where t is the test statistic

$\bar{x}_1$ is the mean of the areas of the *Bergenia* leaf discs after six weeks

$\bar{x}_2$ is the mean of the areas of the *Brassica* leaf discs after six weeks

s_1 is the standard deviation of the areas of the *Bergenia* leaf discs after six weeks

s_2 is the standard deviation of the areas of the *Brassica* leaf discs after six weeks

n_1 is the number of samples of *Bergenia* leaf discs

n_2 is the number of samples of *Brassica* leaf discs

$$s = \sqrt{\frac{\sum(x - \bar{x})^2}{n - 1}}$$

PROCEDURE

Calculate the standard deviation for the *Bergenia* discs (s_1).

1. Find $\sum(x_1 - \bar{x}_1)^2 = 183\,750$

2. Find s_1 where $s_1 = \sqrt{\dfrac{\sum(x_1 - \bar{x}_1)^2}{n_1 - 1}}$

   where $n_1 - 1 = 29$

   $= \sqrt{\dfrac{183\,750}{29}}$

   $= \sqrt{6336.2069}$

   $= 79.6003$

Calculate the standard deviation for the *Brassica* discs (s_2).

3. Find $\sum(x_2 - \bar{x}_2)^2 = 223\,850$

4. Find s_2 where $s_2 = \sqrt{\dfrac{(x_2 - \bar{x}_2)^2}{n_2 - 1}}$

   where $n_2 - 1 = 29$

   $= \sqrt{\dfrac{223\,850}{29}}$

   $= \sqrt{7718.9655}$

   $= 87.8576$

5. Find t where $t = \dfrac{|\bar{x}_1 - \bar{x}_2|}{\sqrt{\dfrac{s_1^2}{n_1} + \dfrac{s_2^2}{n_2}}}$

   $= \dfrac{265 - 340}{\sqrt{\dfrac{79.6003^2}{30} + \dfrac{87.8576^2}{30}}}$

   $= \dfrac{75}{\sqrt{\dfrac{6336.2069}{30} + \dfrac{7718.9655}{30}}}$

   $= \dfrac{75}{\sqrt{211.2069 + 257.2988}}$

   $= \dfrac{75}{\sqrt{468.5057}}$

   $= \dfrac{75}{21.6450}$

   $t = 3.47$ (correct to two decimal places)

Note – the vertical lines indicate that a positive difference between the means must be taken, i.e. +75 not −75.

6. Degrees of freedom $v = n_1 + n_2 - 2$;
   $v = 30 + 30 - 2 = 58$
   If this degree of freedom is not tabulated, take the next value lower than that which is tabulated. In this case take $v = 40$.

7. Look up the critical value of t at the 5% level for this number of degrees of freedom. If t is greater than this critical value, reject the null hypothesis. In this case $t = 3.47$ which is greater than the critical value of 2.021 at the 5% level. The null hypothesis is rejected at 5% level; $p < 1\%$. There is a significant difference in rate of decomposition at $p < 1\%$. The small mean area of the *Bergenia* leaves after decomposition shows they decompose faster than the *Brassica* leaves.

Calculate the standard deviation for the *Brassica* discs (s_2).

Rounding up and down

Do not round up or down until the last step in the calculation. Numbers below 5 are rounded down. Numbers 5 and above are rounded up, e.g. 532.5199235 rounded up correct to two decimal places = 532.52; 6074.321356 rounded down correct to two decimal places = 6074.321; significant figures are counted from the first non-zero figure to the left. 70432.191 rounded down to three significant figures is 70400; 8.558 rounded up to two significant figures is 8.6.

Summary

To find t, find standard deviation for each set of data using the formula:

$$s = \sqrt{\frac{\sum(x - \bar{x})^2}{n - 1}}$$

Assuming that the data come from populations sharing normal distributions with equal variances, t can be found using the formula:

$$t = \frac{|\bar{x}_1 - \bar{x}_2|}{\sqrt{\left(\frac{s_1^2}{n_1} + \frac{s_2^2}{n_2}\right)}}$$

If t is greater than the critical value at 5% level, reject the null hypothesis.

The Chi-Squared (χ^2) Test for Association

When to use this test

- to test for an association between two or more sets of measurements, each at the categorical level.
- provided that each measurement is independent of the others.
- provided that the expected frequencies are larger than five.

Example **7A**	*To test if two long-winged* Drosophila *are heterozygous for wing length*

METHOD

The offspring of a cross between these two flies was 53 long-winged flies and 11 short-winged flies. If both flies were heterozygous for wing length, we should expect a 3:1 ratio of long-winged to short-winged flies. The χ^2 test is used to see if the ratio is close enough to 3:1 to support the hypothesis that both parents are heterozygous for wing length.

FORMULA

$$\chi^2 = \frac{\Sigma(O - E)^2}{E}$$

χ^2 is the test statistic

O are the observed frequencies

E are the expected frequencies

PRESENTATION OF DATA

Table 7.1 Offspring of a cross between two red-eyed *Drosophila*

Numbers of offspring from the cross

Long-winged flies	Short-winged flies	Total
53	11	64

PROCEDURE

1. Tabulate the expected and observed numbers of each eye colour.

   Expected number is the expected fraction × total population.

Expected fraction of long-winged flies is $\frac{3}{4}$.

Expected number of long-winged flies is $\frac{3}{4} \times 64 = 48$.

Expected number of short-winged flies is $\frac{1}{4} \times 64 = 16$.

Wing length of offspring	Numbers of offspring Observed number (O)	Expected number (E)
Long-winged	53	48
Short-winged	11	16

2. Calculate the difference between observed and expected numbers (O − E).

   Long-winged offspring (O − E) $= 53 − 48$
   $= 5$

   Short-winged offspring (O − E) $= 11 − 16$
   $= −5$

3. Square the difference between observed and expected (O − E)².

   Long-winged offspring (O − E)² $= 5^2$
   $= 25$

   Short-winged offspring (O − E)² $= 5^2$
   $= 25$

4. Divide (O − E)² by the expected number E.

   Long-winged offspring $\dfrac{(O − E)^2}{E}$ $= \dfrac{25}{48}$

   Short-winged offspring $\dfrac{(O − E)^2}{E}$ $= \dfrac{25}{16}$

5. Add the values for $\dfrac{(O − E)^2}{E}$ together to find χ^2

   $$\chi^2 = \frac{25}{48} + \frac{25}{16}$$
   $$= 2.08$$

6. Degrees of freedom $v = c − 1$, where c is the number of categories.
   $v = 1$

7. Look up the critical value of χ^2 at the 5% level for these degrees of freedom. If it is greater than or equal to this value, reject the null hypothesis. That is, there is a significant difference between the observed and expected values.

   In this example $\chi^2 = 2.08$ which is less than the critical value of 3.84 at the 5% level ($p = 0.05$). The null hypothesis is accepted. There is no significant difference between the observed and expected numbers.

   The results support the hypothesis that the long-winged *Drosophila* are heterozygous for wing length.

Use of contingency tables in the χ^2 test

Example 7B

Experiment to see if there is an association between bluebell and pink campion in 200 m lengths of roadside verges in the Yorkshire Dales

METHOD
The presence or absence of each species was noted and recorded in each 200 m stretch of roadside verge.

PRESENTATION OF DATA

	Number of 200 m lengths of verge
Both bluebell and pink campion present	72
Bluebell only	14
Pink campion only	22
Neither present	106

PROCEDURE

1. Arrange these numbers in a contingency table:

2. Use χ^2 test to see if there is an association between the bluebell and the pink campion.

	Bluebell present	Bluebell absent	Row total	
Pink campion present	72	22	94	
Pink campion absent	14	106	120	
Column total:	86	128	214	Grand total

FORMULA

$$\chi^2 = \frac{\Sigma(O - E)^2}{E}$$

Assuming there is no association, find expected frequency for each value

$$= \frac{\text{column total} \times \text{row total}}{\text{grand total}}$$

Expected frequency of bluebell with pink campion

$$= \frac{\begin{array}{c}\text{total number of} \\ \text{samples with} \\ \text{bluebell present}\end{array} \times \begin{array}{c}\text{total number of} \\ \text{samples with} \\ \text{with pink campion present}\end{array}}{\text{total number of samples}}$$

$$= \frac{(72 + 14) \times (72 + 22)}{214} = \frac{86 \times 94}{214} = 37.776$$

Expected frequency of bluebell without pink campion

$$= \frac{\begin{array}{c}\text{total number of} \\ \text{samples with} \\ \text{bluebell present}\end{array} \times \begin{array}{c}\text{total number of} \\ \text{samples without} \\ \text{pink campion present}\end{array}}{\text{total number of samples}}$$

$$= \frac{86 \times (14 + 106)}{214} = \frac{86 \times 120}{214} = 48.224$$

Expected frequency of pink campion without bluebell

$$= \frac{\begin{array}{c}\text{total number of} \\ \text{samples with pink} \\ \text{campion present}\end{array} \times \begin{array}{c}\text{number of} \\ \text{samples without} \\ \text{bluebell present}\end{array}}{\text{total number of samples}}$$

$$= \frac{(72 + 22) \times (22 + 106)}{214} = \frac{94 \times 128}{214} = 56.224$$

Expected frequency of neither pink campion nor bluebell

$$= \frac{\begin{array}{c}\text{total number of} \\ \text{samples with pink} \\ \text{campion absent}\end{array} \times \begin{array}{c}\text{number of} \\ \text{samples without} \\ \text{bluebell absent}\end{array}}{\text{total number of samples}}$$

$$= \frac{(14 + 106) \times (22 + 106)}{214} = \frac{120 \times 128}{214}$$

$$= 71.776$$

3. Tabulate observed and expected values:

$$\chi^2 = \frac{\Sigma(O - E)^2}{E}$$

$$= 92.443$$

	Observed O	Expected E	Observed minus expected (O − E)	(O − E)²	$\frac{(O - E)^2}{E}$
Bluebell and pink campion	72	37.776	34.224	1171.28	31.005
Bluebell only	14	48.224	−34.224	1171.28	24.288
Pink campion only	22	56.224	−34.224	1171.28	20.832
Neither bluebell nor pink campion	106	71.776	34.224	1171.28	16.318
					92.443

4. Degrees of freedom = (number of rows in contingency table minus one) × (number of columns in the table minus one)

   = $(2 - 1) \times (2 - 1)$

   = 1

5. Look up the critical value of χ^2 in the statistical table for one degree of freedom at the 5%

significance level. It is 3.841 which is less than this value of 92.443. The null hypothesis is rejected. There is an association between the bluebell and pink campion at the 5% significance level.

The greater number of lengths of verge with both species than lengths with only one of the two species shows that it is a positive association; bluebell can be expected to be found with the pink campion.

Summary of χ^2 test

To find χ^2, tabulate the data:

Category	Observed number O	Expected number E	(O − E)	(O − E)²	$\frac{(O - E)^2}{E}$
long-winged flies	53	48	5	25	25/48
short-winged flies	11	16	−5	25	25/16

χ^2 is the sum of $\dfrac{(O - E)^2}{E}$

$$= \frac{25}{48} + \frac{25}{16}$$

$$= 2.08$$

Degrees of freedom $v = c - 1$, where c is the number of categories.

If χ^2 is equal to or greater than the critical value at $p = 0.05$, reject the null hypothesis at the 5% significance level.

The *t*-test for Matched Samples

When to use this test

- to test for a significant difference between means of two samples.
- using two matched samples.
- for interval measurements.
- for small samples; that is less than about 30 measurements with a minimum of six to ten. Strictly, it is the population of differences between paired measurements that need to be normally distributed.
- assuming that the measurements show a normal distribution.

Example 2A | *To compare the insulating properties of polystyrene granules and beads*

METHOD

Hot water was poured into a calorimeter enclosed by polystyrene beads. The time taken for the water to cool from 90 °C to 80 °C was measured. The experiment was repeated using polystyrene granules instead of beads using the same calorimeter and volume of water. The experiment was repeated eight times.

PRESENTATION OF DATA

Tabulate the following for each replicate:
- time taken for beads to cool from 90 °C to 80 °C
- time taken for granules to cool from 90 °C to 80 °C
- difference between cooling time for beads and granules
- difference between cooling time for beads and granules squared.

Table 8.1 Cooling time using polystyrene beads and granules

Replicates	Time taken to cool from 90 °C to 80 °C /s		difference x	difference squared x^2
	beads	granules		
A	312	305	−7	49
B	291	281	−10	100
C	293	297	+4	16
D	315	314	−1	1
E	241	263	+22	484
F	317	309	−8	64
G	325	331	+6	36
H	321	324	+3	9
I	330	323	−7	49
			$\sum x = 2$	$\sum x^2 = 808$

Construct a bar graph of the cooling times with the beads and granules on the same *x*-axis.

INTERPRETATION OF DATA

Find the test statistic *t* using the formula:

$$t = \frac{\bar{x}\sqrt{n}}{s}$$

$\bar{x}$ is the mean of the difference in cooling time.
n is the sample size (repeated experiments or replicates).
s is the standard deviation for the difference in cooling times.

Find *s* from the formula:

$$s = \sqrt{\frac{\sum x^2 - \frac{(\sum x)^2}{n}}{n - 1}}$$

PROCEDURE

1. Tabulate your results as shown in Table 8.1 with each matched pair on the same line in columns (2) and (3).

2. Find $\bar{x}$ from $\frac{\sum x}{n}$

$$\sum x = 2, n = 9$$
$$\frac{\sum x}{n} = \frac{2}{9}$$
$$\bar{x} = 0.22$$

3. Find $\sum x^2$.

$$\sum x^2 = 808$$

4. Find $(\sum x)^2$.

$$(\sum x)^2 = 2^2$$
$$= 4$$

5. Find $\frac{(\sum x)^2}{n}$

$$\frac{(\sum x)^2}{n} = \frac{4}{9}$$
$$= 0.44$$

6. Find $n - 1$

$$n - 1 = 8$$

7. Find *s*, where $s = \sqrt{\frac{\sum x^2 - \frac{(\sum x)^2}{n}}{n - 1}}$

$$s = \sqrt{\frac{808 - 0.44}{8}}$$
$$s = \sqrt{\frac{807.56}{8}}$$
$$= \sqrt{100.95}$$
$$s = 10.05$$

8. Find $\sqrt{n}$

$$\sqrt{n} = \sqrt{9}$$
$$= 3$$

9. Find *t* where $t = \frac{\bar{x}\sqrt{n}}{s}$

$$t = \frac{0.22 \times 3}{10.05}$$
$$= \frac{0.66}{10.05}$$
$$t = 0.066$$

10. Calculate the number of degrees of freedom *v*, where $v = n - 1$. $v = 8$.

11. Look up the critical value of *t* at the 5% significance level for these degrees of freedom. If *t* is greater than this value, there is a significant difference at $p = 0.05$. The null hypothesis would be rejected.

   In this example, $t = 0.066$. This is less than the critical value of 2.306 at a significance level of 5%. The null hypothesis is accepted. There is no significant difference in insulation properties.

Summary

Find the standard deviation s for the difference in the cooling times from the formula:

$$s = \sqrt{\dfrac{\Sigma x^2 - \dfrac{(\Sigma x)^2}{n}}{n - 1}}$$

Find the test statistic t from the formula:

$$t = \dfrac{\bar{x}\sqrt{n}}{s}$$

If t is less than the critical value at $p = 0.05$, the null hypothesis is accepted.

The Mann-Whitney *U* test

When to use this test

- to test for a significant difference between two samples.
- for ordinal data, i.e. data arranged in order of magnitude or interval data.
- samples must have same shaped distribution but not necessarily a normal distribution.
- can be used for *small* samples provided both samples have more than one measurement and at least one sample has five or more measurements.
- compares the medians of two samples.

Example 9A

To see if there is a significant difference in growth between seedlings with and without a cotyledon removed
Example 1: If neither sample is greater than 20

METHOD

Twenty runner bean seeds were sown in the same plot of land. When the cotyledons (the first seed leaves) had expanded, one cotyledon was removed from each alternate seedling. The other ten seedlings were left intact. After four weeks, all the seedlings were removed, washed and dried in an oven at 100 °C for 24 hours. Each seedling was weighed to find its dry mass.

PRESENTATION OF DATA

Tabulate the dry mass of each seedling.

Construct a histogram to show the frequency of the masses of the two groups of seedlings.

Table 9.1 Dry mass of runner bean seedlings with and without a cotyledon removed

Dry mass of runner bean seedlings after 4 weeks/g

Sample A cotyledon removed	Sample B no cotyledon removed
12	13
13	14
12	16
11	12
15	11
16	14
10	15
13	11
13	10
14	14

INTERPRETATION

Rank the data for each treatment of the seedlings. Find the test statistic U.

PROCEDURE

1. Rank the data in the samples in increasing size.

←smaller										larger→	
Sample A	10		11		12	12	13	13	13		14	15	16
Sample B	10	11		11		12			13		14	14	14	15	16

2. For each measurement in sample B, count how many in sample A are smaller. Write down this score.

If measurements in sample A and B are the same, count a score of $\frac{1}{2}$.

Measurement of sample B	10	11	11	12	13	14	14	14	15	16
Rank of sample B	$\frac{1}{2}$	$1\frac{1}{2}$	$1\frac{1}{2}$	3	$5\frac{1}{2}$	$7\frac{1}{2}$	$7\frac{1}{2}$	$7\frac{1}{2}$	$8\frac{1}{2}$	$9\frac{1}{2}$

U^A = sum of scores of sample B

$U^A = 52\frac{1}{2}$

3. For each measurement in sample A, count how many in sample B are smaller as in (2).

Measurement of sample A	10	11	12	12	13	13	13	14	15	16
Rank of sample A	$\frac{1}{2}$	2	$3\frac{1}{2}$	$3\frac{1}{2}$	$4\frac{1}{2}$	$4\frac{1}{2}$	$4\frac{1}{2}$	$6\frac{1}{2}$	$8\frac{1}{2}$	$9\frac{1}{2}$

Sum of scores of sample A $= U^B$

$U^B = 47\frac{1}{2}$

4. Check that your calculations are correct.

n_A = number of seedlings in sample A

n_B = number of seedlings in sample B

$U_A + U_B = N_A \times N_B$

i.e. $52\frac{1}{2} + 47\frac{1}{2} = 10 \times 10$

$100 = 100$

5. Refer to table of critical values of U at the 5% significance levels.

Take the smallest value of U_A and U_B as the test statistic U.

6. Read the sample sizes from the top and left hand columns of the table to find the critical value.

Critical value for two samples of ten measurements each is 23.

7. If the test statistic U is less than or equal to the critical value, the null hypothesis is rejected, i.e. there is a significant difference between the two samples at the 5% level.

In this case U is $47\frac{1}{2}$ – that is more than 23 – the null hypothesis is accepted. There is **not** a significant difference between the two samples at the 5% level. Removing a cotyledon has no significant effect on the growth of the seedlings after four weeks.

Example 9B | *Example 2: If one sample is greater than 20*

METHOD

Another student removed a cotyledon from 25 seedlings and left 15 intact. As one sample is greater than 20, you can use the Mann-Whitney Z test. A test statistic Z has to be found using the formula:

$$Z = \frac{U - (N_A N_B/2)}{\sqrt{N_A N_B (N_A + N_B + 1)/12}}$$

PROCEDURE

The test statistic U was calculated as in Example 9A.

$U = 148$

$N_A = 15$, $N_B = 25$

$$Z = \frac{148 - (15 \times \frac{25}{2})}{\sqrt{15 \times 25 \times (15 + 25 + 1)/12}}$$

$$= \frac{-39.50}{35.79}$$

$= 1.1037$

$= 1.10$ correct to two decimal places.

Critical value for Z for three significance levels.

Significance level	5%	1%	0.1%
Critical value for Z	1.96	2.58	3.29

INTERPRETATION OF DATA

Reject the null hypothesis at a given level of significance only if the absolute value of your calculated Z (ignoring any minus signs) is greater than or equal to the critical value.

In this example, $Z = 1.75$ which is smaller than 1.96 so the null hypothesis must be accepted at the 5% significance level, i.e. there is no significant difference at the 5% level.

Summary

To find the test statistic, U, score the values for sample A against sample B. Add the values together to give U_A. Score the values for sample B against sample A. Add the values together to give U_B.

Take the smaller value as the test statistic U.

If U is less than or equal to the critical value at the 5% level, reject the null hypothesis.

If one sample size is above 20, find the test statistic Z using the formula:

$$Z = \frac{U - (N_A N_B/2)}{\sqrt{N_A N_B (N_A + N_B + 1)/12}}$$

If Z is greater than or equal to the critical value at the 5% level, reject the null hypothesis.

The Spearman Rank Correlation Coefficient

When to use this test

- to see if there is an association between two sets of measurements that are at the ordinal or interval level.
- when you have samples containing at least seven pairs of measurements.
- when the measurements are reasonably scattered, and do not have a U-shaped or inverted U-shaped relationship with one another.

Three correlation scattergrams where the Spearman test cannot be used

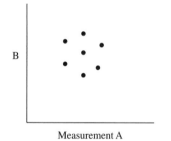

Measurement A

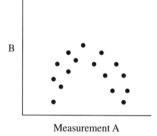

Measurement A

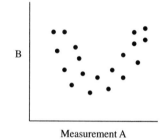

Measurement A

Example 10A — *To see if there is a correlation between the lengths and masses of runner beans*

METHOD

A random sample of eight beans was taken. The maximum length of each pod was measured. Each pod with its enclosed seeds was weighed.

PRESENTATION OF DATA

Tabulate:
- the maximum length of each pod
- the rank of each length. That is, arrange the lengths in order of magnitude. Give the smallest the rank of one. If two measurements

are the same, give them both the average of their rank
- the mass of each pod with its enclosed seeds
- the rank of each mass
- the difference between the ranks of length and mass
- the difference between the ranks of length and mass squared.

Construct a scattergram to show the association between length and mass.

Table 10.1 Length and mass of runner bean pods

Bean number	Length/cm	Rank of length	Mass/g	Rank of mass	Difference in rank D	Difference squared D^2
1	23.6	4	22.1	3	1	1
2	24.5	6	24.0	6	0	0
3	22.1	2	23.8	5	−3	9
4	25.6	7.5	24.3	7	0.5	0.25
5	22.9	3	21.7	2	1	1
6	25.6	7.5	28.9	8	−0.5	0.25
7	20.3	1	21.0	1	0	0
8	24.0	5	23.1	4	1	1
					$\sum D = 0$	$\sum D^2 = 12.5$

INTERPRETATION OF DATA

To see how confident you can be that there is a correlation between maximum length and mass of runner beans, find the test statistic r_S using the formula:

$$r_S = 1 - \frac{6\sum D^2}{n(n^2 - 1)}$$

where D is the difference between ranks and n is the number of pairs of measurements.

PROCEDURE

1. Tabulate the differences in rank for length and mass (D).

2. Square each difference (D^2).

3. Find sum of $D(\sum D)$. If $\sum D$ does not equal 0, there is an error.

4. Find $\sum D^2$

   $\sum D^2 = 12.5$

5. Find n

   $n = 8$

6. Find $n(n^2 - 1)$

   $$\begin{aligned} n(n^2 - 1) &= 8 \times (64 - 1) \\ &= 8 \times 63 \\ &= 504 \end{aligned}$$

7. Find $\dfrac{6\sum D^2}{n(n^2 - 1)}$

   $$\frac{6\sum D^2}{n(n^2 - 1)} = \frac{6 \times 12.5}{504}$$

   $$= 0.149$$

8. Find $1 - \dfrac{6\sum D^2}{n(n^2 - 1)}$

   $$1 - \frac{6\sum D^2}{n(n^2 - 1)} = 0.851$$

9.
   $$r_S = 1 - \frac{6\sum D^2}{n(n^2 - 1)}$$

   $$r_S = 0.851$$

10. Look up the critical value of r_S at the 5% level for the number of pairs of observations n. Take the absolute value of r_S ignoring whether it is positive or negative. If your value of r_S is greater than or equal to the critical value, reject the null hypothesis.

In this example, the value of $r_S = 0.851$ is greater than the critical value at the 5% level of 0.738. The null hypothesis of no correlation at the 5% level is rejected.

The critical value at the 1% level ($p = 0.01$) is 0.881 and 0.833 at the 2% level ($p = 0.02$). The null hypothesis of no correlation is rejected at the 2% level of significance but accepted at the 1% level. ($p < 0.02$ but $p > 0.01$).

There is more than 98% probability of a correlation between the two measurements.

As the line of best fit in the scattergram inclines upwards from zero, there is a positive correlation between length and mass of the bean pods. That is, as the length of the bean pod increases, the mass increases.

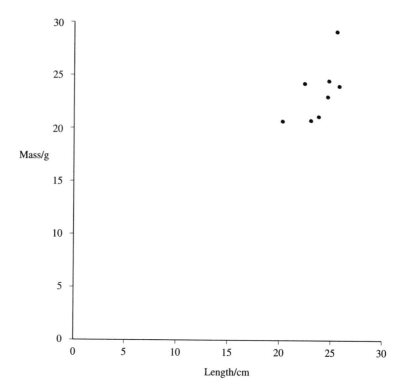

Figure 10.1 *Scattergram to show the relationship between length and mass of runner bean pods.*

Summary

$$r_S = 1 - \frac{6\Sigma D^2}{n(n^2 - 1)}$$

D is the difference in rank between a pair of measurements.
ΣD^2 is the sum of the squares of all the differences.
n is the number of pairs of measurements.
If r_S is greater than the critical value at the 5% level, reject the null hypothesis of no correlation.

If there is a correlation, if the line of best fit of the scattergram inclines upwards, it is positive. If it inclines downwards, it is a negative correlation.

11

The Wilcoxon Matched Pairs Test

When to use this test

- use to test for a difference between population medians.
- provided that you have two matched samples.
- provided that you have interval data.
- provided that the differences between matched measurements are symmetrically distributed.
- do not use if the number of non-zero differences between matched measurements is less than 5 or greater than 30 as the test becomes laborious with larger sample sizes. There are quicker tests that can be used with large samples.

Example 11A | *To compare two red food colouring agents for colour retention in various solutions after various heat treatments*

METHOD

Two red colouring agents, A and B, were subjected to various treatments. The colour retention was estimated visually on a ten point scale.

PRESENTATION OF DATA

Tabulate:
- the treatment of A and B
- the colour retention of A and B on a ten point scale

- the difference in colour retention between A and B
- the rank order of differences in colour retention between A and B.

INTERPRETATION OF DATA

1. Find the differences between A and B.
2. Rank the differences between A and B, ignoring + and − signs. Where the difference is the same, give an average rank, e.g. +1 and −1 have an average rank of 1.5. If two

Table 11.1 Table of colour retention of Agents A and B after various treatments

Colour retention on a ten point scale

Solution or treatment	Agent A	Agent B	Difference between A and B (B − A)	Rank order of differences
1	4	7	3	5.5
2	4	6	2	3.5
3	5	9	4	8
4	1	5	4	8
5	3	6	3	5.5
6	3	3	0	−
7	5	4	−1	1.5
8	6	7	1	1.5
9	6	4	−2	3.5
10	3	7	4	8

matched measurements have the same value (i.e. difference = 0) remove them from the calculations.

3. Divide the ranks into two groups; those associated with positive and those associated with negative differences.

Rank of positive differences	Rank of negative differences
5.5	−3.5
3.5	−1.5
8	
8	
5.5	
1.5	
$\underline{8}$	
40	$\underline{}$
	5

4. Add up the ranks in each of the two groups.

$R^+ = 40.0$

$R^- = 5$

5. Check the accuracy of your calculations.

$R^+ + R^- = \frac{1}{2}N_D(N_D + 1)$

where N_D is number of differences that are not 0.

$N_D = 9$

$40 + 5 = \frac{1}{2}9(9 + 1)$

$45 = 45$

6. Choose the smaller of R^+ and R^- and let this be the test statistic W.

$W = 5$

7. Look up the critical value of W at the 5% level for the value of N_D – in this case $N_D = 9$; the critical value is 5. Reject the null hypothesis if W is less than the critical value at the 5% level. W is the same as the critical value and so the null hypothesis is accepted.

There is no significant difference between the two agents at the 5% level.

Summary

Record the differences in the measurements of the matched pairs. Rank these differences, taking the smallest difference as one. Ignore pairs with zero differences.

Find the sum of the positive differences and the sum of the negative differences. Take the smallest sum as the test statistic W. Reject the null hypothesis if W is less than the critical value at the 5% level.

Worksheets on Statistics

We have discussed why we use statistics and how we choose a suitable test. It is a good idea to see if you can do a statistical analysis on some of the following experiments.

Testing the strength of textiles

In an experiment to investigate the effect on fabric strength of an extra cleaning ingredient in a washing powder, a piece of fabric was cut in half; one half was washed with the standard powder 'Klene' and the other half was washed with the powder with the added ingredient 'Extra-Klene'. The strength of the two fabrics was then compared by cutting identical strips narrowing in the middle to a width of 1 cm. The strips were clamped vertically. Slotted weights were then attached to the lower end until the fabric tore into two. The smallest mass causing tearing was recorded in the table below.

Experiment number	Smallest mass causing tearing with washing powder used/g Klene (A)	Extra-Klene (B)	Difference between Klene and Extra-Klene (A − B)	Difference² (A − B)²
	x_1	x_2	x	x^2
1	745	745	0	0
2	740	760	−20	400
3	770	740	30	900
4	750	760	−10	100
5	720	770	−50	2500
6	745	760	−15	225
7	740	760	−20	400
8	745	745	0	0
9	745	775	−30	900
10	740	760	−20	400

Exercises

A1. What is the sample size (n) for each treatment?
(i)   Sample size for Klene (n_1)
(ii) Sample size for Extra-Klene (n_2)

A2. What is the mean of the mass causing tearing after using:
(i)   Klene (x_1)
(ii) Extra-Klene (x_2)?

A3. Use the formula below to work out the standard deviation in the mass causing tearing for each washing powder:

$$s = \sqrt{\frac{\Sigma(x - \bar{x})^2}{n - 1}}$$

where s is the standard deviation
$(x - \bar{x})$ is the difference from the mean
n is the sample size

A4. What sort of data has been collected in this experiment – interval, ordinal or categorical?

A5. Has the experiment been done on matched pairs or unmatched samples?

A6. Is the experiment looking for a difference or an association?

A7. Which statistical test would you use to see if there is a significant difference in strength of the fabric after the different treatments?

A8. What is the null hypothesis in this experiment?

A9. What is the alternative hypothesis?

A10. Look up the formula for the test you have chosen. Use this formula to find out if there is a statistically significant difference between the two treatments at the 5% significance level ($p = 0.05$). Show your working.

   (i) What is the value of your test statistic?

   (ii) Use the statistical tables to find the critical value of the test statistic at a probability level of 5%.

   (iii) Do you accept or reject the null hypothesis?

A11. What is the probability of the difference occurring by chance?

Height and weight of children in different socio-economic groups

The height and weight of two classes of eight-year-olds were measured. The socio-economic group of each child was recorded based on the occupation of the parents.

Farfield Primary School			St. Swithin's First School		
Girls			Girls		
Height/cm	Weight/kg	Socio-economic group 1 to 5	Height/cm	Weight/kg	Socio-economic group 1 to 5
117	22	5	116	18	4
118	23	5	121	22	1
114	20	3	117	21	2
114	20	–	111	19	3
118	21	3	118	26	5
114	23	–	118	22	–
121	21	3	123	25	3
114	24	4	120	21	2
116	20	–	115	16	2
122	20	3	107	17	3
117	22	–	121	25	3
112	25	3	106	19	2
108	20	3	117	20	3
117	16	–	126	23	3
110	20	3	119	22	2
			119	22	2

| Farfield Primary School | | | St. Swithin's First School | | |
| Boys | | | Boys | | |
Height/cm	Weight/kg	Socio-economic group 1 to 5	Height/cm	Weight/kg	Socio-economic group 1 to 5
118	20	5	122	21	3
122	20	4	129	29	3
115	19	–	113	20	2
112	20	3	112	20	2
116	23	3	131	24	3
125	21	4	116	21	3
128	21	4	114	20	2
122	16	5	123	24	5
131	26	4	112	19	2
118	19	3	121	23	2
102	14	3	113	20	5
117	22	3	115	21	2
117	23	–	117	20	5
			110	19	5

– indicates it was difficult to establish the socio-economic group owing to unemployment

Exercises

B1. What is the sample size (n) of the whole class of:
  (i)   Farfield School
  (ii) St Swithin's School?

B2. What is the mean height of
  (i)   Girls of Farfield School
  (ii) Boys of Farfield School
  (iii) Girls of St Swithin's School
  (iv) Boys of St Swithin's School?

B3. Use the formula below to calculate the standard deviation for the height of
  (i)   all the girls
  (ii) all the boys

$$s = \sqrt{\frac{\Sigma(x - \bar{x})^2}{n - 1}}$$

where s is the standard deviation
$(x - \bar{x})$ is the difference of each measurement from the mean
Σ is 'the sum of'

B4. Comment on the standard deviations of the girls and the boys.

B5. What is the null hypothesis concerning the difference in height of eight-year-old boys and eight-year-old girls?

B6. Choose a statistical test to test this hypothesis bearing in mind:
  (i)   Are you looking for a difference or an association?
  (ii) Is the data interval, ordinal or categorical?
  (iii) Has the investigation been done on matched pairs or unmatched samples?
  (iv) Are the data normally distributed with roughly equal variance?

B7. Look up the formula for the test you have chosen. Use this formula to find out if there is a statistically significant difference between the height of the boys and the girls at the 5% significance level.

B8. Is the null hypothesis supported or rejected?

B9. Use the same formula to find out if there is a statistically significant difference in the heights of all the eight-year-old children measured in Farfield School and St Swithin's School.

B10. What is the probability of the difference in height of the children in the two schools being due to chance?

B11. Choose a statistical test to see if there is an association between height and weight in the eight-year-olds measured in Farfield School bearing in mind:

(i) Are you looking for a difference or an association?

(ii) Is the data interval, ordinal or categorical?

(iii) What is the size of the sample?

B12. Look up the formula for the test. Use the formula to see if there is an association between height and weight. What is the probability of a correlation between height and weight?

B13. What is the null hypothesis in considering the difference in socio-economic groups of the two schools?

B14. Choose a test to see if there is a significant difference in the socio-economic groups of the girls of the two schools. Use the same criteria as in B11 to choose your test. Look up the formula.

Is there a significant difference at the 5% level of significance? Show your working.

Examination Questions

1 Ten children were timed performing a standard task under two different conditions as follows.

(i) With an audience of ten people.
(ii) With no audience.

The results are shown in the table below.

Child	Time to perform task/s	
	Audience	**No audience**
1	70	65
2	42	40
3	67	69
4	36	20
5	47	40
6	61	66
7	54	49
8	22	20
9	49	50
10	57	46

2 The table below shows the number of daisy plants in several one-metre squared samples taken from two different types of grassland, a lawn and a meadow. The lawn was mown regularly once a week and the meadow was continuously grazed by cattle.

The hypothesis was formulated that the children were slower when there was an audience. The details below give some of the values that had to be calculated in order to carry out Student's t-test.

x is the difference in time each child took to perform the task

$\bar{x}$ is the mean difference in time taken by the children

$$\bar{x}^2 = 16$$
$$\sum x^2 = 514$$
$$t \text{ value} = 2.017$$

For a sample number of ten, the critical value of t at the 5% significance level is 2.262.

a (i) How would you have found $\bar{x}$, the mean difference in time taken by the children?
   (ii) How would you have found the value of $\sum x^2$?
b What is meant by a 5% significance level?
c What did the t-test establish about the children's performance under the two different conditions? Explain how you arrived at your answer.

A-level London Specimen Paper Human Biology 1990 Paper 2W

Grassland	Number of daisies in each sample								
Lawn	24	40	44	60	52	76	40	18	24
Meadow	16	20	12	20	12	32	28		

The data were analysed using a Mann-Whitney U test to test the null hypothesis that there was no difference in the population densities of the daisies.

a Arrange the data in rank order above and below the line given in a form suitable for analysis by a Mann-Whitney U test. Then

write in the rank above and below each value. The first few have been done for you. (No. lawn is the number of daisies per sample in the lawn and No. meadow is the number of daisies per sample in the meadow.)

Rank (R_1)				4		
No. lawn				18,		

No. meadow	12,	12,	16,		20,	20,
Rank (R_2)	1.5	1.5	3		5.5	5.5

b The U values can be calculated using the formulae given below

$$U_1 = n_1 \times n_2 + \tfrac{1}{2}n_2(n_2 + 1) - \Sigma R_2$$
$$U_2 = n_1 \times n_2 + \tfrac{1}{2}n_1(n_1 + 1) - \Sigma R_1$$

where n_1 and n_2 are the number of samples looked at.
ΣR_1 and ΣR_2 are the sums of the ranks.
The U value for the lawn is 55 (U_1).
Calculate the U value for the meadow (U_2).
Show your working.

c (i) For this investigation the critical value of U at the 5% significance level is 12. Which value of U would you take to determine the significance of these results?

(ii) Do the results enable you to accept the null hypothesis? Explain your answer.

d Why was the Mann-Whitney U test used in this investigation rather than any other statistical test?

e Explain the effect of mowing on the diversity of species in grassland.

A-level London Biology Paper 1 January 1993

---

3 Read the account of the investigation and answer the questions which follow.

The following naturalistic experiment was carried out on the effects of television viewing on the incidence of prosocial (altruistic or helping) behaviour in children. The experimenters carried out the study in two townships in the USA, one of which could receive television broadcasts and the other – in an isolated rural community – which could not. The researchers were particularly interested in

the impact on the television viewing children of a particular programme called *Rocky Road* which was made especially to help children's prosocial and intellectual development. The programme featured a mixture of animated cartoons and sequences involving interaction between children, adults and pretend animals, which aimed to make learning fun, and showed helping and sharing behaviour in a very positive way.

After a viewing period of six months for the television group, the researchers compared the play patterns of the two groups of children in playground situations, timing over a given period their amount of play in three predetermined categories: aggressive, neutral and prosocial. To improve inter-rater reliability the observers had all been trained in the standardised usage of the classification system before they began their observations of the children taking part in the experiment. Sixty-four five-year-old children were observed in the television township, and 58 five-year-olds were observed in the rural township.

In the statistical treatment the experimenters compared the two groups of children on the amount of time spent engaged in prosocial behaviour. An independent t test was used to analyse the results and test the null hypothesis that there will be no significant difference in the amount of altruistic play behaviour displayed by the two groups. The t value obtained from the results was 2.60 and the experimenters consulted an appropriate table of critical values, the relevant part of which is shown below.

Note: For independent t test
$df = (N_a - 1) + (N_b - 1)$

Level of significance for two-tailed test				
df	.10	.05	.01	.001
40	1.684	2.021	2.704	3.551
60	1.671	2.000	2.660	3.460
120	1.658	1.980	2.617	3.373
∞	1.645	1.960	2.576	3.291

a State an appropriate two-tailed experimental hypothesis for this investigation.

b Identify the independent and dependent variables in the experiment.

c What is a naturalistic experiment?

d Name one difficulty or limitation of the naturalistic experiment and explain how it may have affected this investigation or the conclusions that can be drawn from it.

e What is meant by 'inter-rater reliability' (line 26)?

f How might the play observers have been 'trained in the standardised usage of the classification system' (lines 27–28)?

g Briefly describe how the study might have been carried out in a laboratory setting.

h State one advantage and one disadvantage of this transfer to the laboratory.

i Give two reasons why the independent t test was an appropriate test to analyse the results.

j Which non-parametric test might alternatively have been used?

k What does 'df' stand for?

l State whether the null hypothesis would have been accepted or rejected and explain your reasoning for this decision.

A-level AEB Psychology June 1990 Paper 1

---

4 Read the account of the investigation below and answer the questions which follow.

A social psychologist decided to investigate the contention that the more often we come into contact with a person the more likely we are to like them. She set up the following independent groups design experiment using 180 college students. Subjects were told that the purpose of the experiment was for them to pass judgement on a series of different types of cheese. As part of the experiment they came into contact with another 'student' who was apparently helping the experimenter carry out the study. The 'helper' was actually a confederate and was carefully trained so that he always behaved in the same way during the experiment and did not alter or adjust his behaviour with different subjects. Each of the 180 subjects were allocated to one of four conditions which determined how many contacts each had with the 'helper': one, two, four or ten contacts. After they had finished rating the cheeses each subject was asked to evaluate the 'helper' (so that the experimenter 'could decide whether to use him again in the future'). Each subject was given three choices about their attitude towards the 'helper':

a liked him a lot;

b liked him a little;

c was neutral towards him or did not have a judgement.

The results from the experiment are summarised below.

	no judgement/ neutral	liked him a little	liked him a lot
One contact	24	16	5
Two contacts	16	22	9
Four contacts	13	20	12
Ten contacts	8	18	17

The experimenter analysed the data using a χ^2 test and found that $\chi^2 = 17.16$. The appropriate line from the statistical interpretation tables she used was as follows:

	levels of significance			
df	0.10	0.05	0.01	0.001
6	10.65	12.59	16.81	22.46

a State the null hypothesis for this study.

b What were the independent variable (IV) and dependent variable (DV) in the study?

c What is an independent groups design?

d Explain why a repeated measures design would have been inappropriate in this study.

e Why was it important that the confederate 'helper' was trained to behave in a consistent manner with each subject?

f Explain how the nature of the sample of subjects used might have affected the generalisability of the results. Suggest how this might have been improved.

g What level of measurement was used in this study?

h Is χ^2 a parametric or non-parametric test?

i State three criteria which must be satisfied before a parametric test can be used.

j What does 'df' stand for?

k Explain whether or not the null hypothesis would have been rejected and why.

A-level AEB Psychology June 1989 Paper 1

5 *Angina pectoris* is a severe pain in the muscles of the heart. The table below shows the time that two pain killing drugs (P and Q) were effective for a small sample of eight patients with *Angina pectoris*.

Patient	Time drugs effective/ hours		Difference in effectiveness Q − P	
	P	Q	x	x^2
1	3.2	3.8	0.6	0.36
2	1.6	1.0	−0.6	0.36
3	5.7	8.4	2.7	7.29
4	2.8	3.6	0.8	0.64
5	5.5	5.0	−0.5	0.25
6	1.2	3.5	2.3	5.29
7	6.1	7.3	1.2	1.44
8	2.9	4.8	1.9	3.61
Mean	3.62	4.67	1.05	

A *t*-test was carried out to determine whether the difference in effectiveness was significant at the 5 per cent level.

The formula used for the *t*-test was

$$t = \frac{\bar{x}\sqrt{(n-1)}}{s}$$

where $\bar{x}$ is the mean of the difference in effectiveness

n is the number of patients tested

s is found from the formula

$$s^2 = \frac{\Sigma x^2}{n} - \bar{x}^2.$$

a (i) Calculate the values of $\bar{x}^2$ and Σx^2.

(ii) Use your values from (i) to calculate the value of s. Show your working.

(iii) Use information in the table and your value from (ii) to calculate the value of t. Show your working.

b (i) A statistical table showed that significance at the 5 per cent level with seven degrees of freedom required a *t*-value of at least 2.365.

What does this tell you about the difference in effectiveness of the two drugs as pain killers for *Angina pectoris*?

(ii) Give the formula from which the degrees of freedom were calculated.

A-level London Specimen Paper Human Biology 1990 Paper 3

6 The diagram below shows the seaweed, *Fucus vesiculosus*, held to a rock surface by its holdfast.

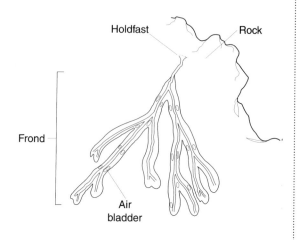

In a study of variation in two populations of this seaweed, the mean holdfast diameter in a sheltered area (P) of a shore was compared with that in an area exposed to heavy wave action (Q).

The table opposite gives the results from ten seaweeds collected at random in each area of the shore. A *t*-test was carried out to determine whether the difference in mean holdfast diameters was significant at the 5 per cent level.

Diameter of holdfast/cm			
Sample from sheltered shore P	Sample from exposed shore Q	Difference in diameters Q − P x	x^2
0.7	1.0	0.3	0.09
0.5	1.0	0.5	0.25
0.7	0.9	0.2	0.04
0.5	0.8	0.3	0.09
1.0	0.7	−0.3	0.09
0.2	0.5	0.3	0.09
0.6	1.0	0.4	0.16
0.7	0.5	−0.2	0.04
0.4	0.6	0.2	0.04
0.4	0.7	0.3	0.09
Mean 0.57	Mean 0.77	Mean 0.2	

The formula used for the t-test was

$$t = \frac{\sqrt{\bar{x}(n-1)}}{s}$$

where $\bar{x}$ is the mean of the differences in diameters

n is the number of samples collected on each shore

s is found from the formula

$$s^2 = \frac{\Sigma x^2}{n} - \bar{x}^2.$$

a (i) Calculate the values of $\bar{x}^2$ and Σx^2.

(ii) Use your values from (i) to calculate the value of s. Show your working.

A-level London Specimen Paper Biology 1990 Paper 3

---

7 It has been suggested that, in humans, certain eye colours and hair colours are often inherited together. The table below shows the observed numbers (O) and the expected numbers (E) for each combination of eye and hair colour found in a sample of 130 British men.

Eye colour	Hair colour						Observed totals
	Fair		Brown		Black		
	O	E	O	E	O	E	
Blue	65	53.3	26	32.0	8	13.7	99
Brown	5	16.7	16	10.0	10	4.3	31
Observed totals	70		42		18		130

The null hypothesis states that there is no link between the inheritance of eye colour and hair colour.

a (i) Test the null hypothesis by means of a χ^2 test.

Fill in the values of (O − E) and (O − E)^2 in the table below.

Combination of eye and hair colour	(O − E)	(O − E)^2
Blue eyes with fair hair		
Blue eyes with brown hair		
Blue eyes with black hair		
Brown eyes with fair hair		
Brown eyes with brown hair		
Brown eyes with black hair		

(ii) Use the formula given below to calculate the value of χ^2. Show your working.

$$\chi^2 = \Sigma \frac{(O - E)^2}{E}$$

(iii) How many degrees of freedom are shown in this investigation? Explain how you arrived at your answer.

b For this number of degrees of freedom, χ^2 values corresponding to important values of P are as follows.

Value of P	0.99	0.95	0.05	0.01	0.001
Value of χ^2	0.020	0.103	5.991	9.210	13.820

What conclusions can be drawn from this χ^2 test concerning the inheritance of eye colour and hair colour?

A-level London Biology June 1992 Paper 3

Glossary of Statistical Terms

A **confidence limit** in data showing normal distribution is defined as a given number of standard deviations either side of the mean. About 95% of the values fall within a confidence limit of plus or minus two standard deviations of the mean.

Critical value is the value of the test statistic at a particular probability, usually 5%. If the value of the test statistic is greater than the critical value, then the results can be said to be statistically significant at the 5% level ($p < 0.05$). That means there is less than 5% probability of these results occurring by chance.

Data are a collection of facts.

Categorical (nominal) data is data of different types which cannot be arranged in order of size, e.g. brown mice and white mice in genetics experiments.

Interval data is measured in units where it is possible to say exactly how much greater one measurement is than another, e.g. height in humans in cm.

Ordinal data is the arrangement of data in rank order, i.e. in order of size but where it is not possible to say exactly how much greater one measurement is than another, e.g. measuring intensity of colour in leaves as pale green, medium green and dark green.

Degrees of freedom are the number of independent observations on which the test is based. It is denoted by the letter v.

$$v = n - 1$$

where n is the number of observations in the sample.

In considering the difference between two samples:

$$v = n_1 + n_2 - 2$$

In testing for an association between two or more sets each at the categorical level $v = c - 1$, where c = number of categories.

The degrees of freedom need to be known to use the statistical tables.

Hypothesis is a suggested explanation of something based on the facts.

Null hypothesis is that there is no difference between the experiment and the control, e.g. 'there is no statistically significant change in heart rate after ingesting caffeine'.

Alternative hypothesis is that there is a difference between the experiment and control, e.g. 'there is a statistically significant change in heart rate after ingesting caffeine'.

Two-sided hypothesis predicts there will be difference but does not state the direction of the difference, e.g. 'there is a statistically significant change in heart rate after ingesting caffeine'.

One-sided hypothesis predicts the direction of the difference, e.g. 'there is a statistically significant increase in heart rate after ingesting caffeine'.

Matched pairs are used in an experiment in which the same or a closely related subject or object is used for each condition. The heart rate of the same water flea is measured at different temperatures. There is a unique link between a measurement in one set of data and a measurement in the other set.

Related measures is an experiment done with matched pairs.

Unmatched samples are samples or subjects used in an experiment in which different samples or subjects are used in each condition. The heart rate of one water flea is measured at 10 °C and the heart rate of another water flea is measured at 20 °C. There is no suggestion of pairing between measurements of the two samples.

Independent measures is an experiment designed with unmatched subjects.

Parametric test is a statistical test with interval data, normal distribution of data with equal variability, e.g. the t-test for unmatched subjects or matched pairs.

Mean – arithmetic mean is the sum of the measurements divided by the number of measurements. It can only be used for interval data. It is shown by $\bar{x}$ (pronounced x bar).

$$\bar{x} = \frac{\Sigma x}{n}$$

where n is number of measurements

  Σ is 'sum of'

x are the measurements.

Median is the value above which half of the measurements in the sample lie and below which the other half lie. In the mass of leaves/g:

15.5, 15.5, 15.5, 18.5, 18.6, 19.0, 19.4, 19.7, 19.9

The median is 18.6. If there was another reading above 18.6 g, the median would be mid-way between 18.6 and 19.0 = 18.8 g.

Mode is the most frequent measurement.

Probability is the chance of an event occurring. It is denoted by the letter p. $p = 0.05$ means that there is a 5 in 100 chance of an event occurring.

Normal distribution is a distribution of measurements which fall symmetrically round the mean in a bell-shaped curve. The mean, median and mode coincide (see Fig. 3.2).

Skewed distribution is one in which the mode does not coincide with the mean (see Fig. 3.1).

Bimodal distribution is one in which there are two modes. The human population might show this as the average height of males is greater than females.

Sample is the set of measurements taken.

Sample range is the difference between the highest and lowest measurement.

Subject is the organism or object being experimented on, e.g. *Daphnia*.

Standard deviation is a sort of average distance of the measurements from the sample mean. It is the square root of the variance denoted by s.

Sample standard deviation is the standard deviation of a small sample.

$$s = \sqrt{\frac{\Sigma(x - \bar{x})^2}{(n - 1)}}$$

Population standard deviation is the standard deviation of the whole population or a very large sample. The population is all possible measurements.

$$s = \frac{\Sigma(x - \bar{x})^2}{n}$$

Standard error is the difference between sample means denoted by SE_D.

$$SE_D = \sqrt{\frac{s_1^2}{n_1} + \frac{s_2^2}{n_2}}$$

Statistical significance is a measure of the probability of the results occurring by chance. To show statistical significance, the test statistic must be:

greater than the critical value at $p = 0.05$	equal to or less than the critical value at $p = 0.05$
t-test χ^2 test Spearman rank correlation coefficient Mann-Whitney Z test	Mann-Whitney U test Wilcoxon matched-pairs test

Variance is a measure of the average distance from the mean. It is denoted by s^2.

$$s^2 = \frac{\Sigma(x - \bar{x})^2}{(n - 1)}$$

Variable is a measurement capable of changing.

Independent variable is the measurement or situation chosen by the experimenter. In the effect of temperature on heart rate of *Daphnia*, the temperature is the independent variable (IV).

Dependent variable is the effect of the IV being measured, e.g. the heart rate of *Daphnia*.

Confounding variables are variables which cannot be controlled. The swimming speed of *Daphnia* cannot be controlled.

Appendix

Answers to Worksheets

A1.   (i)   10
      (ii)   10

A2.   (i) $\bar{x}_1$ is 744 g
      (ii) $\bar{x}_2$ is 757.5 g

A3. s_1 for Klene is 12.20
     s_2 for Extra-Klene is 11.13

A4.   Interval data

A5.   Matched pairs

A6.   Difference

A7. t-test for matched pairs

A8.   There is no difference in the effect of the two washing powders on the strength of the fabric

A9.   There is a difference in the effect of the two washing powders on the strength of the fabric

A10. $t = \dfrac{\bar{x}\sqrt{n}}{s}$ where $s = \sqrt{\dfrac{\sum x^2 - \dfrac{(\sum x)^2}{n}}{n-1}}$

    $\bar{x} = 13.5$
    $s = 21.088$
    (i) $t = 2.024$
    (ii)   Critical value at $p = 0.05$ and $v = 9$ is 2.262
    (iii) Accept the null hypothesis

A11. Probability of the difference occurring by chance is greater than 5% but less than 10%

B1.   Sample of the whole class at Farfield School (n_1) is 28
    Sample of the whole class at St. Swithin's School (n_2) is 30

B2.   (i)   115.467 cm
      (ii)   117.125 cm
      (iii) 118.692 cm
      (iv) 117.714 cm

B3.   (i)   4.66
      (ii)   6.83

B4.   The standard deviation of the boys is greater than that of the girls.

B5.   There is no significant difference between the height of the eight-year-old boys and girls.

B6.   (i)   Difference
      (ii)   Interval
      (iii) Unmatched samples
      (iv) Data show normal distribution with unequal variance.

B7.   Mann-Whitney Z test

$$Z = \frac{U - (N_A N_B/2)}{\sqrt{N_A N_B (N_A + N_B + 1)}_{12}}$$

    $Z = 0.756$
    Degrees of freedom $v = 58 - 2$
                      $= 56$

Critical value for Z at p at 0.05 is 1.96.

Z is less than the critical value so there is no significant difference at the 5% significance level.

B8. The null hypothesis is supported.

B9. t is 0.048. There is no significant difference between the heights of the children at the two schools.

B10. The probability of the difference in the heights of the children in the two schools being due to chance is greater than 20%.

B11. (i) Association

(ii) Interval

(iii) 28

B12. Spearman's rank

$$r_S = 1 - \frac{6\Sigma D^2}{n(n^2 - 1)}$$

$$= 0.1337$$

Critical value at 5% level is 0.377

r_S is less than the critical value so there is no correlation between height and weight.

The probability of a correlation between height and weight is less than 1%.

B13. There is no difference between the socio-economic groups of the pupils of the two schools.

B14. (i) Difference

(ii) Ordinal

Mann-Whitney U test

U_A is $37\frac{1}{2}$; U_B is $112\frac{1}{2}$

U_A is the test statistic. It is less than the critical value of 39 at the 5% significance level.

1. The Mann-Whitney U test

Criticial values of U at the 5% level. Reject your null hypothesis at the 5% level if your value of U is less than or equal to the tabulated value.

n_1/n_2	1	2	3	4	5	6	7	8	9	10	11	12	13	14	15	16	17	18	19	20
1	–	–	–	–	–	–	–	–	–	–	–	–	–	–	–	–	–	–	–	–
2	–	–	–	–	–	–	–	0	0	0	0	1	1	1	1	1	2	2	2	2
3	–	–	–	–	0	1	1	2	2	3	3	4	4	5	5	6	6	7	7	8
4	–	–	–	0	1	2	3	4	4	5	6	7	8	9	10	11	11	12	13	13
5	–	–	0	1	2	3	5	6	7	8	9	11	12	13	14	15	17	18	19	20
6	–	–	1	2	3	5	6	8	10	11	13	14	16	17	19	21	22	24	25	27
7	–	–	1	3	5	6	8	10	12	14	16	18	20	22	24	26	28	30	32	34
8	–	0	2	4	6	8	10	13	15	17	19	22	24	26	29	31	34	36	38	41
9	–	0	2	4	7	10	12	15	17	20	23	26	28	31	34	37	39	42	45	48
10	–	0	3	5	8	11	14	17	20	23	26	29	33	36	39	42	45	48	52	55
11	–	0	3	6	9	13	16	19	23	26	30	33	37	40	44	47	51	55	58	62
12	–	1	4	7	11	14	18	22	26	29	33	37	41	45	49	53	57	61	65	69
13	–	1	4	8	12	16	20	24	28	33	37	41	45	50	54	59	63	67	72	76
14	–	1	5	9	13	17	22	26	31	36	40	45	50	55	59	64	67	74	78	83
15	–	1	5	10	14	19	24	29	34	39	44	49	54	59	64	70	75	80	85	90
20	–	2	8	13	20	27	34	41	48	55	62	69	76	83	90	98	105	112	119	127

Dashes indicate no decision possible at the stated level of significance.

n_1/n_2	1	2	3	4	5	6	7	8	9	10	11	12	13	14	15	16	17	18	19	20
16	–	1	6	11	15	21	26	31	37	42	47	53	59	64	70	75	81	86	92	98
17	–	2	6	11	17	22	28	34	39	45	51	57	63	67	75	81	87	93	99	105
18	–	2	7	12	18	24	30	36	42	48	55	61	67	74	80	86	93	99	106	112
19	–	2	7	13	19	25	32	38	45	52	58	65	72	78	85	92	99	106	113	119
20	–	2	8	13	20	27	34	41	48	55	62	69	76	83	90	98	105	112	119	127

Dashes indicate no decision possible at the stated level of significance.

2. The Wilcoxon matched-pairs test

Critical values of W at various significance levels. Reject null hypothesis if your value W is less than the tabulated value at the chosen significance level, for the number of non-zero differences N_D.

	significance level			
N_D	10%	5%	2%	1%
5	0	–	–	–
6	2	0	–	–
7	3	2	0	–
8	5	3	1	0
9	8	5	3	1
10	10	8	5	3
11	13	10	7	5
12	17	13	9	7
13	21	17	12	9
14	25	21	15	12
15	30	25	19	15
16	35	29	23	19
17	41	34	27	23
18	47	40	32	27
19	53	46	37	32
20	60	52	43	37
21	67	58	49	42
22	75	65	55	48
23	83	73	62	54
24	91	81	69	61
25	100	89	76	68
26	110	98	84	75
27	119	107	92	83
28	130	116	101	91
29	140	126	110	100
30	151	137	120	109

3. The *t* test for matched and unmatched samples

Critical values of *t* at various significance levels. Reject the null hypothesis if your value of *t* is larger than the tabulated value at the chosen significance level, for the calculated number of degrees of freedom.

degrees of freedom	significance level					
	20%	10%	5%	2%	1%	0.1%
1	3.078	6.314	12.706	31.821	63.657	636.619
2	1.886	2.920	4.303	6.965	9.925	31.598
3	1.638	2.353	3.182	4.541	5.841	12.941
4	1.533	2.132	2.776	3.747	4.604	8.610
5	1.476	2.015	2.571	3.365	4.032	6.859
6	1.440	1.943	2.447	3.143	3.707	5.959
7	1.415	1.895	2.365	2.998	3.499	5.405
8	1.397	1.860	2.306	2.896	3.355	5.041
9	1.383	1.833	2.262	2.821	3.250	4.781
10	1.372	1.812	2.228	2.764	3.169	4.587
11	1.363	1.796	2.201	2.718	3.106	4.437
12	1.356	1.782	2.179	2.681	3.055	4.318
13	1.350	1.771	2.160	2.650	3.012	4.221
14	1.345	1.761	2.145	2.624	2.977	4.140
15	1.341	1.753	2.131	2.602	2.947	4.073
16	1.337	1.746	2.120	2.583	2.921	4.015
17	1.333	1.740	2.110	2.567	2.898	3.965
18	1.330	1.734	2.101	2.552	2.878	3.922
19	1.328	1.729	2.093	2.539	2.861	3.883
20	1.325	1.725	2.086	2.528	2.845	3.850
21	1.323	1.721	2.080	2.518	2.831	3.819
22	1.321	1.717	2.074	2.508	2.819	3.792
23	1.319	1.714	2.069	2.500	2.807	3.767
24	1.318	1.711	2.064	2.492	2.797	3.745
25	1.316	1.708	2.060	2.485	2.787	3.725
26	1.315	1.706	2.056	2.479	2.779	3.707
27	1.314	1.703	2.052	2.473	2.771	3.690
28	1.313	1.701	2.048	2.467	2.763	3.674
29	1.311	1.699	2.043	2.462	2.756	3.659
30	1.310	1.697	2.042	2.457	2.750	3.646
40	1.303	1.684	2.021	2.423	2.704	3.551
60	1.296	1.671	2.000	2.390	2.660	3.460
120	1.289	1.658	1.980	2.158	2.617	3.373
∞	1.282	1.645	1.960	2.326	2.576	3.291

For larger sample sizes, the degrees of freedom are infinity (i.e. $v = \infty$) and *t* is the same as Z.

4. The χ^2 test for association

Critical values of χ^2 at various significance levels. Reject the null hypothesis if your value of χ^2 is bigger than the tabulated value at the chosen significance level, for the calculated number of degrees of freedom.

degrees of freedom	significance level												
	99%	98%	95%	90%	80%	70%	50%	30%	20%	10%	5%	2%	1%
1	0.000157	0.000628	0.00393	0.0158	0.0642	0.148	0.455	1.074	1.642	2.706	3.841	5.412	6.635
2	0.0201	0.0404	0.103	0.211	0.446	0.713	1.386	2.408	3.219	4.605	5.991	7.824	9.210
3	0.115	0.185	0.352	0.584	1.005	1.424	2.366	3.665	4.642	6.251	7.815	9.837	11.341
4	0.297	0.429	0.711	1.064	1.649	2.195	3.357	4.878	5.989	7.779	9.488	11.668	13.277
5	0.554	0.752	1.145	1.610	2.343	3.000	4.351	6.064	7.289	9.236	11.070	13.388	15.086
6	0.872	1.134	1.635	2.204	3.070	3.828	5.348	7.231	8.558	10.645	12.592	15.033	16.812
7	1.239	1.564	2.167	2.833	3.822	4.671	6.346	8.383	9.803	12.017	14.067	16.622	18.475
8	1.646	2.032	2.733	3.490	4.594	5.527	7.344	9.524	11.030	13.362	15.507	18.168	20.090
9	2.088	2.532	3.325	4.168	5.380	6.393	8.343	10.656	12.242	14.684	16.919	19.679	21.666
10	2.558	3.059	3.940	4.865	6.179	7.267	9.342	11.781	13.442	15.987	18.307	21.161	23.209
11	3.053	3.609	4.575	5.578	6.989	8.148	10.341	12.899	14.631	17.275	19.675	22.618	24.725
12	3.571	4.178	5.226	6.304	7.807	9.034	11.340	14.011	15.812	18.549	21.026	24.054	26.217
13	4.107	4.765	5.892	7.042	8.634	9.926	12.340	15.119	16.985	19.812	22.362	25.472	27.688
14	4.660	5.368	6.571	7.790	9.467	10.821	13.339	16.222	18.151	21.064	23.685	26.873	29.141
15	5.229	5.985	7.261	8.547	10.307	11.721	14.339	17.322	19.311	22.307	24.996	28.259	30.578
16	5.812	6.614	7.962	9.312	11.152	12.624	15.338	18.418	20.465	23.542	26.296	29.633	32.000
17	6.408	7.255	8.672	10.085	12.002	13.531	16.338	19.511	21.615	24.769	27.587	30.995	33.409
18	7.015	7.906	9.390	10.865	12.857	14.440	17.338	20.601	22.760	25.989	28.869	32.346	34.805
19	7.633	8.567	10.117	11.651	13.716	15.352	18.338	21.689	23.900	27.204	30.144	33.687	36.191
20	8.260	9.237	10.851	12.443	14.578	16.266	19.337	22.775	25.038	28.412	31.410	35.020	37.566
21	8.897	9.915	11.594	13.240	15.445	17.182	20.337	23.858	26.171	29.615	32.671	36.343	38.932
22	9.542	10.600	12.338	14.041	16.314	18.101	21.337	24.939	27.301	30.813	33.924	37.659	40.289
23	10.196	11.293	13.091	14.848	17.187	19.021	22.337	26.018	28.429	32.007	35.172	38.968	41.638
24	10.856	11.992	13.848	15.659	18.062	19.943	23.337	27.096	29.553	33.196	36.415	40.270	42.980
25	11.524	12.697	14.611	16.473	18.940	20.867	24.337	28.172	30.675	34.382	37.652	41.566	44.314
26	12.198	13.409	15.379	17.292	19.820	21.792	25.336	29.246	31.795	35.563	38.885	42.856	45.642
27	12.879	14.125	16.151	18.114	20.703	22.719	26.336	30.319	32.912	36.741	40.113	44.140	46.963
28	13.565	14.847	16.928	18.939	21.588	23.647	27.336	31.391	34.027	37.916	41.337	45.419	48.278
29	14.256	15.574	17.708	19.768	22.475	24.577	28.336	32.461	35.139	39.087	42.557	46.693	49.588
30	14.953	16.306	18.493	20.599	23.364	25.508	29.336	33.530	36.250	40.256	43.773	47.962	50.892

© Chalmers and Parker, *OU Project Guide*, Field Studies Council.

5. The Spearman rank correlation coefficient, r_s

Critical values of r_s at various significance levels. Reject the null hypothesis if your value of r_s is greater than or equal to the tabulated value at the chosen significance level, for your number of pairs, n.

no. of pairs, n	significance level			
	10%	5%	2%	1%
5	0.900	1.000	1.000	–
6	0.829	0.886	0.943	1.000
7	0.714	0.786	0.893	0.929
8	0.643	0.738	0.833	0.881
9	0.600	0.683	0.783	0.833
10	0.564	0.648	0.746	0.794
12	0.506	0.591	0.712	0.777
14	0.456	0.544	0.645	0.715
16	0.425	0.506	0.601	0.665
18	0.399	0.475	0.564	0.625
20	0.377	0.450	0.534	0.591
22	0.359	0.428	0.508	0.562
24	0.343	0.409	0.485	0.537
26	0.329	0.392	0.465	0.515
28	0.317	0.377	0.448	0.496
30	0.306	0.364	0.432	0.478

© Chalmers and Parker, *OU Project Guide*, Field Studies Council.

Bibliography

Bassett et al. (1986). *Statistics Problems and Solutions*. Edward Arnold.

Clarke, G.M. (1980). *Statistics and Experimental Design*. Edward Arnold.

Cadogan, A. and Sutton, R. (1994). *Mathematics for Biologists*. Thomas Nelson, London.

Chalmers, N. and Parker, P. (1989). *The OU Project Guide*. Field Studies Council.

Fowler, J. and Cohen, L. (1990). *Practical Statistics for Field Biology*. Open University Press.

Graham, A. (1988). *Teach yourself Statistics*. Hodder & Stoughton.

Greer, A. (1980). *First Course in Statistics*. Stanley Thornes.

Gonick, L. and Smith, W. (1993). *Cartoon Guide to Statistics*. Harper.

Hayslett, H.T. and Murphy, P. (1983). *Statistics Made Simple*. Heinemann.

Johnson and Kent. (1988). *Elementary Statistics*. Longmans.

Jones, R. (1994). *Statistics Work Book*. Pitman.

Owen, F. and Jones, R. (1994). *Statistics*. Pitman.

Rees, D.G. (1995). *Essential Statistics*. Chapman & Hall.

Rowntree, D. (1981). *Statistics without Tears*. Penguin.

Walker, McLean and Matthew. (1993). *Statistics – a First Course*. Hodder & Stoughton.

Acknowledgements

My thanks to the students of Park Lane College, Leeds and Harrogate Tutorial College for providing the data for most of the investigations in this book, and to Mary Underwood for checking the calculations.

Grateful acknowledgements to FSC publications for permission to reproduce the statistical tables from Chalmers, N. and Parker, P. (1989), *The OU Project Guide*, Field Studies Council.

The publishers would also like to thank the following for permission to reproduce copyright question material:

The University of London Examinations and Assessment Council, and the Associated Examining Board.

The illustrations were drawn by Hugh Neil, and thanks to Robert Bull for additional assistance.

Cover photograph by Jonathan Scott, courtesy of the Telegraph Colour Library.